To Arlene, Mary, Neeta, Esther, Peggy and Roberta,
who have been caregivers to my children, and
whose faithful hours and efforts have permitted
me to reap the joys of science.

—CMY

To all the women of science past and present, who
have struggled, persevered, and achieved, despite
resistance and even hostility from a highly
conservative system.

—CJS

CONTENTS

We must welcome the future remembering that soon it will be the past, and we must respect the past remembering that at one time it was all that was humanly possible.

—George Santayana

PROLOGUE

This book is an attempt to describe today's women scientists, with emphasis on the differences and similarities in their roles, accomplishments, career satisfactions, and attitudes compared with men. It is not a deliberate search for discrimination or unequal treatment, but rather an attempt to achieve a balanced view of today's professional woman and her work environment. To this end, positive as well as negative aspects of careers in science have been included.

The objective is, as the title suggests, to examine the questions "What is the present female experience in science and what is the prognosis for the future?" To meet this objective we have examined the career progressions, attitudes, successes, and seeming failures of women scientists through interviews, small group conferences, questionnaires, seminars, and conversations with colleagues, looking particularly for patterns as well as unique experiences of female professionals.

The sample group consists of science professionals of varying ages, status, and accomplishments, without undue concentration on the highly successful individuals. Samples have been drawn mainly from academia, with emphasis on university faculty members. Industry and government people have a unique set of situations that are not covered in detail in this book. Every attempt has been made to operate from a reasonable statistical base, but the strengths of the book should be in opinions and perceptions, presented without too many preconceived points of view. When an individual opinion is expressed, it will be indicated by quotation marks. When a perception is expressed, whether the perception is true or not, it will be denoted as such in the text.

The book is constructed around discussions of issues that affect women scientists. Those that have been explored include:

- The changing image of women scientists
- The changing work atmosphere
- The increase in relative numbers of female scientists*
- Participation in research groups by women
- Research initiatives introduced by women
- Special challenges associated with supervising female scientist employees
- Special challenges related to interactions with female colleagues
- Special challenges for women bosses
- Special challenges for female agency managers
- Special challenges associated with travel and high-endurance duty

*We will be using "women scientists" and "female scientists" interchangeably.

- Special challenges of women working with women
- Gender equality in pay and promotion

Subsidiary issues that have been included are:

- Careers, marriage, and motherhood
- The role of diplomacy in the success of women scientists
- Taking women scientists seriously
- Female status (single, married, divorced, pregnant, new mother, menopause) as considered by those in decision-making roles
- Harassment of women scientists
- Women's movements and their influences on careers
- "Invisible woman" complex
- "Impostor syndrome"

Special features of the book are:

1. The use of a unique time line approach—a graphic presentation of important personal events in the life of a female scientist—that can be interwoven with career progression.

2. Joint female and male authorship, which permits the expression of perceptions from interviews that may be quite different.

The potential advantage (and some of the risk) of female–male authorship of a book on women scientists is the possibility—no, probability—of varying perspectives on some of the issues that are discussed. In an attempt to illustrate the extent of the challenge, we have cited on the next few pages a number of case histories, which readers may use as a test of their own perceptions.

In fact, the book might be considered merely a collection of two authors' opinions aided by those individuals who have confided in them. Both authors' lives span a period when their scientific community was almost exclusively male, to the present, when a diversified, more integrated community is being heralded. Young women scientists may find the lack of conviction of the authors on some important recently stressed issues disturbing. The authors attribute this difference in intensity of feelings to generational differences.

Some of the issues raised in the case histories will be readdressed and explored in subsequent chapters of the book, together with indications of whatever common ground we have found and even suggestions for future actions.

Case History No. 1: Recruiting

Three finalists—two males and one female—have been identified from an applicant list of more than one hundred for an entry-level position of assistant professor of biochemistry at a major research university. The first male is just completing a three-year postdoctoral appointment with one of the key figures in his specialty area and has lectured in several courses. The second male has been an instructor at a state university, where he has been recognized as an innovative teacher and has published several research papers.

The female has just completed an Ivy League Ph.D.; she has excellent references but no postdoctoral or teaching experience. All three are obviously well qualified for this entry-level position. A faculty review committee has recommended male number two, but the female finalist was selected for the job. What would be your comments on this case?

Case History No. 2:
Tenure Decision for a Feminist Activist

Dr. Laura Harkness is an assistant professor of chemistry scheduled for tenure decision this year. During her five years as a university faculty member she has achieved a reputation as an outspoken activist for women's rights and chairs the women's faculty association. She has done some research in her specialty area, but has published little and has offered what are considered adequate but uninspiring courses, principally at the undergraduate level. Will she get tenure?

Case History No. 3: Leave of Absence

Dr. Marcia Llandgraff is an assistant professor of microbiology at a midsized, privately funded university. She teaches courses at undergraduate and graduate levels and has recently become a mother after a difficult pregnancy. Her courses suffered and her research time was reduced to near zero during her pregnancy. The university has a limited maternity leave policy, but she has requested a year's leave of absence with half salary and no loss of benefits. What course should the university administration follow?

Case History No. 4: Research versus Administration

Dr. Carolyn Schneider is a tenured university professor with large research grants in the developing field of biotechnology. Her grants support several postdocs and a cadre of technicians, assistants, and graduate students. Recently, in a rare contemplative moment, she accepted the fact that she had created her own small high-stress universe, complete with daily crises and staff demands, and that some of the touted joys of science were proving elusive, despite her growing professional eminence. Almost simultaneously with that realization, she was invited to consider a job as vice president for academic affairs at an elite four-year private college in another state. The new position would have prestige, but would effectively end her active research involvement. If you were Dr. Schneider, would you accept?

Case History No. 5: Profession versus the Home

Dr. Maria Long is a junior research scientist in a federal environmental laboratory. She has procured three-year funding for an oceanographic project that she conceived and initiated. She has one technologist working with her on the project. In support of the project, Dr. Long has been granted one month of ship time for years 1 and 3.

Dr. Long gave birth to a daughter four months ago and is scheduled to go to sea next month for one month. She has made preparations to wean her child this month and has found a suitable live-in sitter to cover during her absence. If you were Dr. Long, would you go on the research cruise or get someone else to go in your place?

Case History No. 6: Establishing Selection Criteria

In 1991, the Center for Molecular Studies established a Young Scientist Award and Week-long Forum for high school awardees in response to the need for science and math education reform. Information about the program was circulated through the high school principals and local newspapers. The application form consisted of biographical data and a short essay question, "What is science?" Selection of participants was made by a committee consisting of an institutional representative, the head of the state's high school principals association, the president of the high school science teachers association, and two institutional trustees—all male. Sixteen students were selected. When the disclosure of the successful awardees was made, ten were male and six were female. When confronted with the hope that equal proportions were to be represented, the retort was, "The males were just better candidates." What action should be taken? At what level?

Case History No. 7: Hierarchical Relationships

Christopher Gent is a masters' level research associate who has been employed to work on the grants and contracts of Dr. Ann Maya for the past seven years. They share a pleasant collegial relationship and have numerous joint authored publications. Christopher conducts most of the experimental manipulations; Ann writes most of the proposals and manuscripts. When Christopher disagrees with Ann, his immediate supervisor, he goes above her to her supervisor, the director of the laboratory (male). What action should be taken? What advice should be given? By whom?

Case History No. 8: Sex in the Laboratory

Dr. Joe Sibon and his graduate student, Rhonda Green, enjoyed a close professional relationship, which developed into a close personal relationship. While every effort was made by both parties to behave discreetly in public, they became the subjects of scrutiny. What action should be taken? By whom?

The core of the data on which conclusions in this book have been based consists of results of more than 200 detailed interviews, roughly two-thirds female and one-third male, which followed a prescribed format. Interviewees—male as well as female—included the broadest attainable spectrum of groups—graduate students to emeritus professors—and of professional attainments. One small problem, identified early in the interview phase of the work, was that some women scientists appeared uncomfortable with a male interviewer, and often seemed to be giving the kind of answer the interviewer expected rather than their inner thoughts. This kind of potential bias, based on the perception that women are less candid with men, was reduced by increasing the frequency of reinterviews and decreasing the proportion of female interviews conducted by the male author. Although some reviewers may question the "robustness" of our data on purely technical grounds, we are convinced that the information obtained can be useful and that it represents the views of at least a reasonable segment of the scientific population.

We also want to make it perfectly clear that, except where indicated, all names of scientists, their specialties, and their institutional affiliations have been changed to ensure the anonymity that was guaranteed by us as a condition for our interviews and discussions. The closing paragraphs of any respectable prologue should be devoted to acknowledgments. We agree with this practice, so we thank all the scientists, female and male, who invested substantial amounts of their time in completing our complex questionnaires and in providing useful insights during interviews. We are indebted to all the forever-unnamed women scientists who assisted in the partial reconstruction of the second author, who had much to atone for as a consequence of warped viewpoints and outrageous perspectives on women published in his earlier books about scientists. We also thank our Plenum editor, Linda Greenspan Regan, who has remained convinced of the worth of the project, and who has provided continuing stimulus for its completion. We thank the many participants in feminist activities—especially during the 1960s, 1970s, and 1980s—whose efforts have been reflected in some of the movement toward equal status for women scientists documented in this book. And we are indebted to those who have published results of recent studies and analyses of women scientists, especially J. Barnard, R. K. Merton, J. R. Cole, A. Simeone, H. Zuckerman, S. Weiler, P. Yancy, and V. Gornick, whose findings and conclusions have been referred to repeatedly in this book and have been important in broadening its perspectives.

We invite your constructive comments on the conclusions reached in this volume, since additional evaluations should help broaden the data base that we have examined here.

CHAPTER 1

INTRODUCTION

Status of Women Scientists
in the Twentieth Century

*Optimistic hypotheses about the improving status of
women in science; commonly held beliefs about women
scientists; a plan for examining the current state of affairs.*

In 1975, Harriet Zuckerman and Jonathan R. Cole,[1] two
authors who have published extensively about women scientists, listed a number of widely held beliefs or perceptions
about the historic treatment of women in science and re-
viewed the existing literature as a foundation for their study
of patterns of discrimination. The list of perceptions, derived from a number of surveys and not necessarily en-
dorsed by the authors, can be summarized as follows:

- Women are neither fit for scientific careers nor interested in them.

- Women are discriminated against in admission to graduate school.

- Women are poor risks as graduate students; if they acquire advanced degrees, they then marry, have children, and leave science.

- Women are less likely to have received training at the most distinguished universities.

- Women are less productive scientists than men.

- Women scientists do not receive rewards commensurate with the quality of their work.

- Women receive less informal recognition from colleagues for the quality of their work.

- Women suffer the consequences of accumulative disadvantages during their entire careers.

- Social conditions that have impeded career progress for women in science have not changed materially in the past fifty to seventy-five years.

Intuitively, some of these widely held beliefs, as assembled by Zuckerman and Cole and subsequent observers, seem unnecessarily harsh and inconsistent with available evidence. Some may still be true, at least in part. With certainty, beliefs persist in many instances, despite obvious indications of progress in acceptance of women as integral components of the scientific community.

The perception of a constantly improving role for women scientists deserves periodic reexamination, however, to assess whether or not the reality fits the perception. To guide such an analysis, one approach would be to state a series of hypotheses and then assemble whatever data may seem relevant. We have done that, and our hypotheses are these:

- Women scientists are now in the mainstream of science and are no longer on the fringes.

- Many women scientists now direct scientific research as well as do it. Only the numbers remain disproportionate.

- Relatively few women go into the scientific fields and fewer still, proportionally, end up in academia.

- The requirements for success in science for females are not qualitatively different from those for males—but quantitatively more is expected.

- Principal criteria for success in science are, to women, contributions to knowledge and *recognition* of those contributions—these criteria often ranking above financial rewards.

- The study of women in science is a study of *survivors* of a male-dominated system, since a majority of those who once thought they might be scientists have given up early—some even as undergraduates.

- Women tend to congregate in the less quantitative sciences (psychology, sociology, biology, and anthropology) rather than in the more quantitative sciences (mathematics, physics, chemistry, and engineering).

- The proportion of women decreases as one moves up the hierarchy from entry level to the top; the proportion of women decreases as one moves from local, state, national, to international participation.

- With increasing empowerment of women (numbers, acceptance), the criteria for evaluating effective scientists are beginning to be refocused. Emphasis is moving away from power and monographs toward developing human resources, scholarly interaction, and sharing of ideas and data.

- The subject matter of choice by women scientists and the manner in which they pursue their science, more frequently than with male scientists, dwell on investigation of interaction, cooperation, and mutualism as opposed to competition.

Reasonable doubts can be raised immediately about these hypotheses. Take the matter of women scientists as survivors. A logical question might be: "Are they then like survivors of a smallpox epidemic—scarred, resistant, and toughened by the experience?" An intuitive answer might be: "Some are, but most are tolerant, perceptive, even idealistic (but their stance may become negative and hard when pressed about lingering gender-related inequities)." Or take the issue of the increase in numbers of women as directors of research projects. A logical question would be: "There may be some, but what proportion of science managers are female?" The intuitive answer might be: "The proportion is still remarkably small since, because of earlier inequalities in the system of science, fewer women occupy the pool of senior positions from which managers are selected."

To escape from the subjectivity that is inherent in perceptions and intuitions, we will try to provide a dispassionate analysis of case histories of female and male scientists. The approach is not perfect and certainly not unique, but it should lead to an approximation of reality—to a portrayal of the current status of women professionals in a highly conservative system still dominated by men. This is the direction we want to pursue in this book.

It would seem only fair to point out that an extensive literature exists about women in science—most of it written by women. Some of it contains detailed case histories of successful women scientists; some of it is written from feminist positions; and some of it is encased in the almost-impenetrable jargon of sociology. We have tried to avoid the multiple pitfalls of an overemphasis on a few successful careers, the subjectivity that is lurking in feminist tracts, and the complex language of the sociologist. As simple biologists, we may be accused of naiveté, narrowness of sample population, or misinterpretation of data, but we have done our best to eliminate biases or intrusions of personal agendas.

With this proclamation of purity of purpose, we have aggregated our analyses under the following major chapter headings:

- career goals of women scientists
- education and training of women scientists
- a time line approach to women's life-styles as career scientists
- employment, underemployment, and unemployment among women scientists
- women in scientific support roles
- women scientists in positions of power and influence
- women scientists as role models and mentors
- mobility among women scientists

- participation by women in the social infrastructure of science
- perceptions and realities
- subtle forms of gender-based discrimination in science
- eras in scientific careers
- a forecast for the future

A logical and almost necessary part of this introductory chapter should be our definitions of terms as they will be used in the book. They are:

chauvinism: invidious attachment or partiality for a group to which one belongs

feminism (as defined by Simeone[2]): "a belief in the equality of women and men, a sense that women are oppressed in today's society, support for women's rights, and a commitment to act in ways which are consistent with these values."

feminist: one who promotes female rights and interests

gender (as distinct from sex): male or female

sex: erotic interaction

sexist: erotic intent or expectation addressed to females as a group; viewing the opposite sex in a stereotypic manner that may not apply

sexual harassment: unwelcome erotic intent forced on an individual; or the continuous intrusion of a sexual component into a professional relationship

A legal definition of sexual harassment might be useful here, too. In the United States, harassment on the basis of sex is a violation of Section 703 of Title VII of the Civil Rights Act of 1964 as amended. It is defined as:

> Unwelcome sexual advances, requests for sexual favors, and other verbal or physical conduct of a sexual nature constitute sexual harassment when (1) submission to such conduct is made either explicitly or implicitly a term or condition of an individual's employment, (2) submission to or rejection of such conduct by an individual is used as the basis for employment decisions affecting such individual, or (3) such conduct has the purpose or effect of unreasonably interfering with an individual's work performance or creating an intimidating, hostile, or offensive working environment.

The terms used in describing the consequences of gender biases in science can have different meanings to different people. To begin with, a dictionary definition of bias is "such prepossession with some object or point of view that the mind does not respond impartially to anything related to the object or point of view." Gender bias would therefore include all the consequences of that prepossession as applied to one gender—as addressed in this book, particularly to women. The consequences or expressions of gender bias can be lumped, for discussion purposes, into three categories: sexual harassment (including its ugly stepchild "gender-based harassment"), gender-based discrimination, and differential professional or social treatment based on gender.

- *Sexual harassment* may be verbal or physical. Verbal forms include sexual innuendo directed toward one person and being propositioned on the job. Physical harassment includes a variety of unwanted physical contacts—touching, kissing, fondling, brushing against, pinching—on and on, and all now illegal. Ranked just below sexual harassment, in our opinion, is an often-unpublicized *gender-based harassment*. At issue here is not so much the erotic as it is the perils of being a woman. Included in this subcategory would be all the overt and often vicious manifestations of male hostility, particularly belittling, devaluing, and ignoring women and their accomplishments. Such attacks extend beyond the subtle devices listed later under the category "differential professional and social treatment." Gender-based harassment can be very direct and meant as a punishment of an individual woman for all the fancied wrongs inflicted on defenseless male scientists.

- *Gender-based discrimination* may be overt or subtle. Overt forms include inequities in hiring practices, promotions, salaries, awards, and tenure decisions. Subtle forms include exclusion, invisibility, and a variety of other kinds of relatively minor slights, some of which intergrade with the third category, "differential professional treatment" (Table 1).

- *Differential professional or social treatment based on gender* is more nebulous and a little harder to encompass. The category would include all the minor annoying, embarrassing, or gratuitous actions directed toward women solely because of their gender: not being taken seriously by male colleagues, accusations of being too vocal or overbearing, being recognized in a scientific forum because of being a woman rather than a scientist, being made to feel like a second-class citizen in the workplace, or being subjected to rumors that a hiring decision was made because of gender.

TABLE 1
PERCENTAGE OF AMERICAN ASTRONOMICAL SOCIETY (AAS)
RESPONDENTS WHO REPORTED WITNESSING OR EXPERIENCING
DISCRIMINATION[a]

	Against women		Against minorities: Percent of minorities
	Percent of women	Percent of men	
General social treatment	50	15	8
Promotions	34	9	10
Accommodations–special circumstances	34	8	0
Pay/fringe benefits	31	8	5
Tenure decisions	26	6	7
Hiring practices	37	14	14
Research opportunities	25	5	10
Accommodations–job mobility	16	3	1
Opportunities to give talks, etc.	18	2	6
Competition for institution's resources	17	3	4
Administrative appointments	14	3	4
Admission of students to graduate programs	17	4	8
Nominations, elective offices, honor societies	13	2	5
Committee assignments	12	2	6
Competition for grants/fellowships	14	2	5
Teaching assignments	10	1	7
Prizes and awards	11	2	6
Access to research facilities	9	2	3

[a]2208 of 5600 AAS members responded; 12.6% of the respondents were women.
From Barbara Spector, Women astronomers say discrimination in field persists, *The Scientist* (1 April 1991), 20.

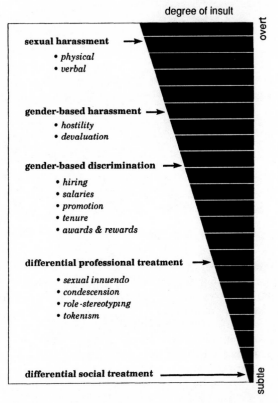

FIGURE 1. Schematic representation of expressions of gender bias in science.

A schematic representation of the degree of insult from various expressions of gender-based bias can be found in Figure 1. Distinctions among these somewhat artificial categories can get fuzzy at times. The perimeters of sexual harassment are not universally agreed upon, and the boundary between differential social treatment and discrimination is not always crisp. For example, gender bias is evident when inappropriate questions about family plans are asked during a job interview, whereas outright discrimination would include being denied the interview altogether because of gender. We probably should say a little more about sexual harassment, beyond defining it, for two important reasons: a surprising proportion of women scientists claim to have been sexually harassed at some point in their careers,[3,4] and the many nuances and interpretations of the term sexual harassment can lead to distorted conclusions about its prevalence and virulence.

To ensure proper consideration, sexual harassment should be differentiated as clearly as possible from other expressions of gender bias. It is direct unwelcome erotic intent or action, aimed at one particular person (usually a woman), and it may be physical or verbal. It can be differentiated (but with some difficulty) from all other sexually oriented actions—innuendo, jokes, disparaging comments—since it is directed at one particular woman and not women as a group. Sexual harassment should be separated from gender-based harassment, which has less erotic content and is essentially verbal abuse heaped on a woman because she is female. Sexual harassment should also be separated from the many other forms of discrimination and differential treatment—professional and social—accorded to women scientists (Table 1). Most of the latter are subtle, but still demeaning and destructive, as we will try to indicate in some of the following chapters.

Each chapter of this book is nominally independent, but each one should contribute information to support or refute the hypotheses already stated and should also contribute to some ephemeral unity—which may not become too apparent until the final pages are read. To provide a semblance of structural cohesiveness, many of the major chapters follow an internal format of perceptions, realities, suggested actions, and conclusions.

We have tried to be diligent about distinguishing the perceptions of others—true or false—from conclusions reached by us. The distinctions can be important in a subject matter area such as this one where viewpoints and opinions can be strongly held and forcefully expressed.

We freely admit that this book is primarily about academic women scientists, principally because of the nature of our interests and sampling. Undoubtedly women scientists in government and industry would have their own stories to tell. They may achieve deserved visibility in someone else's book.

We have stated our objectives and have outlined our plan of action; now on to the results of our investigation of the current status of women in science.

REFERENCES

1. Harriet Zuckerman and Jonathan R. Cole, Women in American science, *Minerva* 13(1), 82–102 (1975).
2. Angela Simeone, *Academic Women: Working toward Equality* (Bergin and Garvey, South Hadley, MA, 1987), 97.
3. Barbara Spector, Women astronomers say discrimination in field persists, *The Scientist* (April 1991), 20–21.
4. Bernice R. Sandler and Roberta M. Hall, *The Campus Climate Revisited: Chilly for Women Faculty, Administrators, and Graduate Students* (Association of American Colleges, Washington, D.C., 1986).

CHAPTER 2

CAREER GOALS OF WOMEN SCIENTISTS

"What do they really want?" Criteria for success in science
from a woman's perspective: institutional recognition—
positions, salaries, and tenure; professional recognition—
prizes, awards, and society offices; gender differences
in productivity.

Some, possibly many, male scientists have wondered in quiet moments, "What do they really want from their careers and from me as a colleague?" (the "they" referring to women scientists). Answers to this unspoken and obviously chauvinistic question are elusive and variable. But examination of the responses of women scientists to questions about reasons for their initial and continuing involvement in science discloses evidence for a triumvirate of primary career goals:

1. Substantive contributions to new knowledge.

2. Peer recognition and respect.

3. Acceptance by "the system" (which includes institutional parity with male scientists in rank, salary, and tenure).

Financial rewards, which once lurked in the background, have become as important to women as to men. However, to women the financial rewards are often viewed as a criterion of others' recognition of their value rather than as a primary career objective. Open discussion of salaries and other personal goals seems most prominent in the age group under 35—the group in which changes in attitudes toward careers can be very clearly identified.

Success is important in any career. Success in science is an ephemeral concept, with complex components and strong disagreements about the relative importance of each component. One reasonable definition of success is that professional state which results from significant contributions to human understanding of natural phenomena, recognized by peers, and internalized by the contributor. Criteria for success in science will vary with individual practitioners, but a common core seems to exist, with principal emphasis on these elements:

- Achievement of respect from colleagues and establishment of credibility among them as a thinker and producer.

- Acceptance into and active participation in the real but ill-defined "in-groups" that exist within any subdiscipline.

- Nomination and election to society offices and long-term participation in society boards of directors or standing committees.

- Recognition by panels of peers in the form of prizes and awards for outstanding accomplishments (in publication, concept development, excellence in teaching, etc.).

- Frequent and persistent invitations to participate in symposia, to chair sessions, and to present lectures or keynote addresses.

- Invitations or nominations to chair national or international meetings, workshops, working groups, conferences, and symposia.

- Invitations to serve on editorial boards of journals and to act as an editor of a scientific journal.

- Invitations to serve on peer review panels, study committees, and evaluation teams.

- Appointment to research administrative posts as group leader, division manager, or laboratory director.

A logical initial question might be: "Faced with such a high degree of individual variability, can any consistent gender-related differences be discerned in how men and women perceive success in science?"

Many women find these criteria of success in science more desirable than the prevailing societal goal of wealth. Many of the items listed refer to power, even if only in the scientific community. To such professionals, significant contributions to the improvement of knowledge can be a criterion of worth that outweighs financial success. The

creativity and mental challenges implicit in scientific research become sources of satisfaction and pleasure not
found in many other occupations. Male or female, they want
innovation, flexibility, and the ability to be creative at work.
These are already—and have always been—strong elements
in any scientific career.

Traditionally, women who succeed in science are those
who mimic men; women who have "identified with the
aggressor," in the words of Vivian Gornick[1] in her book
Women in Science: Portraits from a World in Transition, and have
followed the work and thought patterns of the only ones
doing science (the men). The patterns, as Gornick sees them,
are the outcome of hundreds of years of assimilated male
experience; they include aggressive self-confidence despite
loss, failure, or defeat and the ability to "walk the earth as
though they owned it." Women have heretofore not had
such experiences in science or elsewhere.

Some women scientists refuse to mimic men. Dr. Ce-
leste Morgan, a physical oceanographer, made the following
statement after reading Betty Harragan's book, *Games Mother
Never Taught You*[2]:

> I have just finished reading the book *Games Mother Never Taught
> You* and I am mystified. It is cold, straightforward, and calcu-
> lated. I can see according to the author's prescription, where I
> have gone wrong in my career. I now realize where and when my
> career advancement has been curtailed. I have initiated more
> externally sponsored grant programs than any male in my
> division, yet I have the least power. What is more, I realize that I
> don't want to adopt the characteristics (which I associate as male
> games) necessary to win by the rules that men play. I refuse to be
> the kind of person who steps on others for personal advance-
> ment. Upon reflection, I have a different kind of power, a power
> based on respect by my cherished colleagues and subordinates,
> and recognition outside of my institution.

INSTITUTIONAL RECOGNITION: POSITIONS AND SALARIES OF WOMEN SCIENTISTS

Science, as a remarkably conservative human institution despite its relatively brief history, has typically cast women in supporting roles in which they were subservient to male professionals, usually dreadfully underpaid, and totally unrecognized. Changes have occurred at an accelerating pace in recent decades—changes that will be documented in later chapters of this book. We will in this section be concerned initially with changes in positions within institutional hierarchies and with concomitant changes in financial rewards. Some perceptions that may or may not be true are these:

- Salaries for women scientists are gradually approaching (some would say "inching toward") parity with those for men, but differentials are still common in many institutions.

- To many women scientists, particularly in the over-35 age category, money is less important than the pleasures of involvement in good science—in meaningful research.

- Promotions in rank still do not reflect gender parity, and many women, because of their more recent entry into science in numbers, remain clustered in the lower ranks.

- Persistent imbalances exist in percentages of women in tenured or tenure-track positions.

Some interesting statistics about career advancement that support the last two listed perceptions were discussed in a 1983 symposium on women in science and summarized in the journal *Science* by John T. Bruer.[3] In one 1981 survey of women and men who had received their doctorates during 1970–1974, remarkable gender differences in academic rank and tenure were found (Figures 2 and 3). The understated conclusion from that study was that "Similarities in training and first job experience do not result in comparable careers for men and women."

A more recent study by a professors' union in a large Southeastern state university system reported that, as an average in 1987, women were paid $1325 less than men of the same rank, field, and experience. Other findings of the study, as described by Cathy Shaw in the *Miami Herald*,[4]

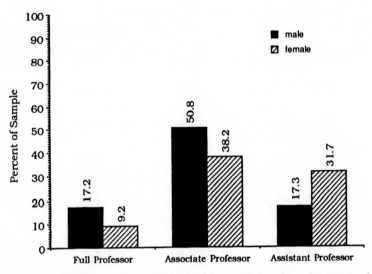

FIGURE 2. Faculty ranks in 1981 of males and females who received science doctorates in 1970–1974. SOURCE: John T. Bruer.[3]

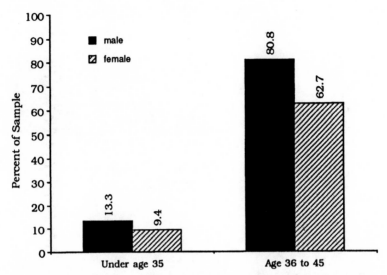

FIGURE 3. Percentages in 1981 of males and females in tenured or tenure track positions (science doctorates received in 1970–1974). SOURCE: John T. Bruer.[3]

were that women tend to be hired in the lowest paid positions in disciplines that are low in pay, that women accept lower starting salaries and rarely ever catch up, and that women work in less prestigious jobs and move more slowly from the lower ranks.

One of the most interesting findings was that in 1987, women faculty members started out at average salaries of $26,432 compared with $34,855 for men. Of course, statistics can't clarify all aspects of any problem. One plausible reason given for the disparity was that women tend to gravitate to the traditionally poorly paid humanities rather than to the better-paid fields of business, science, and engi-

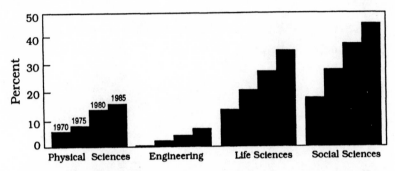

FIGURE 4. Percent of total Ph.D.s granted to females in various scientific disciplines, 1970–1985. SOURCE: Jonathan R. Cole and Harriet Zuckerman, 1987.[5]

neering (Figure 4), where universities must compete with private industry for professional employees.

Another interesting comment made about the study by a teachers' union member, with less supporting information, was that once a profession becomes female-dominated, the average salaries become depressed—and this could be interpreted as a sign of persistent sexism.

Disparities in salaries were clearly identified in earlier studies.[5,6] Differences in 1970 between males and females who received their degrees around 1950 were reported to range from 28% in mathematics to 15% in economics and sociology,[7] but those authors concluded that most of the differences could be attributed to differential productivity, years of service, rank, and specialized training, rather than to overt discrimination. A smaller degree of difference was noted in another earlier study by the National Education Association.[8] Salaries for women scientists were less than those for men by only 5% at the instructor level, 6% at the assistant/associate professor level, and 10% at the full professor level.

These and other studies of gender-related salary differences suggest the existence of only a small amount of discrimination even 20 years ago,[9] and are in accord with more recent surveys, including our own, which suggest that the small salary gap has narrowed still further. What seems to have persisted, however, is the continuing significant relationship between gender and salary at only the highest professional levels, with nonsignificant relationships at lower academic levels. Of course, these are *statistical* relationships, and as Jonathan Cole[6] pointed out, discrimination may still exist in *individual* cases, even though the aggregate trend has been toward equity in salaries.

Any discussions about salaries, especially comparative salaries, invade areas very private to most scientists, so they have to be approached cautiously and the results interpreted conservatively. Responses from a majority (75%) of the women scientists surveyed by us indicated that in their estimation their salaries were equal to or approximately equal to those of male counterparts in the same institution at the same career stage and with similar abilities. Responses could be categorized as follows: not sure/don't know, higher than male, equal, slightly less, or significantly less. A few of the replies revealed degrees of uncertainty because of institutional policies discouraging public release of individual salaries, whereas a number of replies pointed to legal mandates or union agreements specifying equal financial treatment.

Other estimations about comparative salaries, made by women scientists queried by us, are of interest too. Only one respondent stated that her salary was higher than that of male counterparts, but almost 20% stated that their salaries were slightly lower than male counterparts at the same career stage.

The general conclusion that can be drawn from this admittedly limited probe is that in their own perception a majority of women scientists in our survey feel that their salaries are comparable to those of men at the same career stage. This conclusion is decidedly more optimistic than those reached by others less than a decade ago.[3] This disparity may be a consequence of the restricted nature of the population sampled or it may reflect an unrealistic appraisal. Do most women really know the salaries of their male cohorts? We hope that genuine progress toward financial parity for women in academic science is firmly underway.

Tenure

Long a subject of controversy in the United States academic system, tenure has in the past two decades come under increasing scrutiny and is being challenged in many universities. Partly because of restricted financial growth, tenure decisions are approached with extreme caution, and so-called nontenure track appointments (short-term contracts, part-time appointments, research associateships) have become very popular with administrators. Senior faculty members are being induced to accept early retirement, and their courses are then either discontinued or taken over by young nontenured instructors or assistant professors, often on one-, two-, or three-year contracts, thereby creating a lower-cost, easily replaceable (or easily removable) faculty.

In this competitive climate, it is logical to ask, "Are women scientists being forced to assume a disproportionate share of the consequences of this reductionist movement?" Conclusions about this matter that can be drawn from our data are not crisp, except for one. Since more women are at the lower end of the scale in terms of rank, because of their relatively recent entry in numbers, restrictive institutional policies on tenure have disproportionate impact on them as a group, even in the absence of any subtle discriminatory practices by the institution. It must be remembered, too, that the senior faculty involved in tenure decisions is still predominantly male, and some of these people may still harbor well-concealed doubts about women as tenured faculty members or they may feel threatened by the greater proportion of women eligible for and rightfully expecting tenure.

PROFESSIONAL RECOGNITION: PRIZES, AWARDS, AND SOCIETY OFFICES FOR THE WOMAN SCIENTIST

The reward system of science, unlike that of many other professions, has a dual nature. Institutional recognition, principally in the form of advancement in rank and salary, is balanced by recognition from the larger scientific community, especially in the form of invited participation in the major forums of science (national and international conferences and symposia, peer review panels, receipt of prizes and awards, and election to society offices). Women (as well as men) depend on constant reaffirmation from the larger scientific community. How do women scientists fare in these broader arenas, beyond the walls of the employing institutions?

A simplistic answer, as deduced from our data, is that women scientists compete successfully with men and that professional recognition has a rapidly diminishing gender component. Such an answer does not imply that inequalities and discrimination have disappeared altogether—especially among the in-groups and clubs that characterize many specialties—but it does imply that once the limiting attitudes and strictures of local organizations are transcended, rewards tend to be based principally on merit and performance and not on chromosomal complement.

THE ENIGMA OF GENDER DIFFERENCES IN SCIENTIFIC PRODUCTIVITY

Throughout the history of science in the twentieth century, a succession of studies has consistently pointed to a continuing difference in productivity of male and female scientists. Earlier in the century, before World War II, it was possible to explain the disparity between males and females by pointing to the many overt exclusionary and discriminatory practices faced by women scientists—the common perception that women were less capable of creative research, the denial of professional positions, and the greater disparities in salary and rank. The situation during the period from 1945 to 1975 was summarized by Budner and Meyer[10] and Cole.[6] A consistent pattern of significantly greater research productivity and publication by males was reported, despite greater access of females to facilities and funding. Productivity differentials were found to exist and to increase with age, and some hypotheses were proposed by Cole to explain the correlations found:

- Women tended to be employed in primarily teaching institutions, where research productivity was not highly valued, and may, in fact, have been discouraged.

- High research productivity may not have been reflected in promotions and salary increases, so such productivity, in the absence of motivation, declined.

- A greater proportion of women entered science with a primary interest in teaching rather than in research.

- Patterns of socialization and psychological traits of female scientists may have helped explain differences in publication rates.

- A cumulative self-reinforcing disadvantage may have occurred in which women produced less work that made a significant impact in their specialty area, so they were less likely to attend meetings or to be invited to present lectures, to be involved in peer review decisions, or participate in symposia. Some of the consequences were that they were less likely to be chosen for advisory panels or society offices and less likely to receive honors or awards.

- Women may not have been as successful as men in converting the results of their work into other forms of recognition and rewards, either by choice or by societal and institutional constraints.

Reasons for lower productivity of women scientists were also explored by Jonathan Cole and Stephen Cole in an earlier book.[11] Their general conclusion, although stated as a speculation, was that ". . . productivity differences will ultimately be explained by differences in motivation." They pointed out that patterns of early socialization, resulting in lower commitment to careers and reduced drive to achieve success in science, may have contributed to the differences seen. Cole and Cole also suggested the interesting possibility that receipt of the Ph.D. may have had a significantly different meaning for males and females: for men it represented an important "union card" for entry into science, but not an end in itself, whereas for women the Ph.D. may have been viewed as the culmination of a long and difficult journey, with less motivation to charge ahead beyond that point and to produce further results. The consequent lower productivity in the several years following award of the degree may have, in the opinion of those authors, contributed to "accumulative disadvantage" from which women never fully recovered.

The conclusions reached in these studies about underlying reasons for the apparent differences in productivity were reexamined in an excellent paper by Jessie Bernard.[12] She emphasized possible negative effects of caring for two or more children and the reduction in productivity caused by career interruptions for childbearing and rearing. Her opinion on the matter, stated in a footnote, was that ". . . marriage per se may not [hinder scientific productivity] but children often do."

Some other seemingly plausible explanations for the gender disparity in publication of research results have been examined and discarded. Exclusionary practices by journal editors and reviewers have not been identified as problems by female journal editors and associate editors. A 1987 study by Jonathan Cole and Harriet Zuckerman[5] provided convincing evidence that marriage and motherhood did not significantly affect research production by women scientists. Interesting subsidiary findings of that study were that ". . . married women with children publish no fewer papers than married women without children . . ." and that "Women scientists married to scientists publish, on the average, 40 percent more than women who are married to men in other occupations."

The importance of a partner's support to success in science is well illustrated by the comments of Dr. Noellia Gee.

Both of my husbands have been highly supportive, in fact, my most enthusiastic supporters. I married my first husband before we both started graduate school. He was also a chemist. Whenever I would get discouraged and consider quitting, he would convince me to continue. He had unshakable faith in my ability. He also knew what had happened to his mother who did not finish her Ph.D. in chemistry. He was absolutely right and when he died after our 10 years of marriage, I was able to support our three children.

My second husband is not a scientist, but is extremely enthusiastic about my career. He understands its importance to me, but isn't jealous or threatened by it. He is also convinced that I'm better at it than I think I am. When he was offered a far better job than he had, probably the most interesting job he will ever have, he was prepared not to take it because of my career and my recent positive tenure decision at a prestigious university. Fortunately I found an almost equivalent job in the location of his offer, which allowed him to accept the offer and me—although the "trailing spouse"—to flourish. What I found most touching was his genuinely sensitive and equal consideration for my career in our mutual decision.

Productivity throughout scientific careers may be influenced to a marked degree by the existence and quality of a mentor–protégé(e) relationship. An essential component of the relationship was described by Jonathan Cole[6] as "sponsorship"—the complex of activities that provides incipient scientists with crucial advantages in early career development. Included would be informal discussions, examples of research styles, criteria for choices of research problems, invitations to workshops and seminars, actual collaboration in research projects, and placement in optimum first positions. Cole asked the question that is still relevant today: "Is the probability lower for women students to participate in these informal experiences that will eventually be reflected in enhanced scientific productivity and status?" He offered no clear answers, but pointed to real or perceived problems such as close working relationships of male professors and female protégées, which may make male faculty members reluctant to initiate any associations beyond the most formal. This thereby denies women graduate students the many advantages of informal interactions that can be so important to developing scientific careers. He also suggested that at least some of the marked productivity differentials between female and male scientists might have their roots in patterns of sponsorship. We think this suggestion is highly pertinent today.

The relationships of high scientific productivity and recognition, in the form of promotions, salary, and awards, are not simple, nor is the influence of gender. Jonathan Cole[6] found that awards were directly related to scholarly output, regardless of gender, but that ". . . productivity patterns simply will not explain sex [gender] differences in academic rank"—leading him to conclude that extensive gender discrimination in promotion opportunities existed prior to 1975, and remnants have persisted. The world was different then. The reasons for lower scholarly output by women scientists remain elusive, but the reality of the difference remains. Cole felt that this was a critical issue, stating that "To understand sex [gender] differences in science careers, more work must be focused on research productivity" and, further, that ". . . without some reasonable measure of productivity, it is virtually impossible to establish credible estimates of discrimination."

The reality today seems to be that women scientists as a statistical entity produce somewhat less than men, but those individuals who publish results of scholarly work have a better opportunity than ever before for higher faculty positions in major research universities. The quality and quantity of published research results seem critical to establishment of scientific reputations, as reflected in rank and status in the professional community. This point is important since, as Cole[6] emphasized, differential rewards in science may reflect differences in quality of performance rather than gender discrimination, so the measurement of performance should be built into proposed affirmative action plans.

SUGGESTED ACTIONS

Attaining career objectives and achieving success in science result from the confluence of many internal and external factors and are only marginally predictable. Some gender-free approaches to success include:

- Substantial original research in an area of expertise
- Extensive publication in the best journals in that discipline area
- Publication of substantive, authoritative, superbly written reviews and technical books
- Significant contributions to ideas, concepts, analyses, and syntheses in an area of expertise
- Enthusiastic and informed participation in workshops and symposia
- Effective oral presentations in society meetings and at conferences

- Ability to attract good graduate students and to aid in their transformation into scientists

- Effective—even exceptional—teaching in formally scheduled courses, either at undergraduate or graduate levels

- Development and maintenance of close professional and personal relationships with a group of productive colleagues of both genders

- An almost instinctive ability to assume a dominant role in committees, informal workshops, peer review panels, and similar group activities (this stems in part from credibility as a scientist, enthusiasm for the subject matter, and careful consideration of the prerogatives of others)

Armed with a reasonable sampling of these attributes, women scientists have every right to expect institutional and professional rewards commensurate with those accorded to men, and that expectation is being realized to a far greater extent than in the past.

The route to these rewards is rarely if ever easy, however. This reality was well described in great detail by one of our respondents.

In the mid-1970s, Dr. Shana Stone was offered several attractive positions near the end of her postdoctoral fellowship. The institution where she did her postdoctorate wanted to keep her and offered her a tenure track position. A small, nonprofit research laboratory that prided itself on a 50:50 male–female representation at the senior scientist and administrative levels offered her a position as a research scientist. A large private university in a major U.S. city offered her an assistant professorship, tenure track. Her major professor (male) and her postdoc advisor (male) both advised her to go "hard core." "No women in this discipline are at a major university, you will be the first." She made this decision. Although her grant and publication records were stellar for the ensuing seven years at that university, she was denied tenure. Many feel that it was because she was compared to a "super woman" in another discipline at that university who was granted tenure the previous year. In comparison to other males who obtained tenure, she was very bright, yet in comparison to the one female, dim.

She was offered jobs at several prestigious yet less lofty institutions, but she wanted to stay. Depression set in. Female colleagues from other institutions encouraged her to fight back and to appeal the tenure recommendation. She did so, insisting that a female be part of the tenure review committee. Her request was granted and her tenure approved, almost without question. Whether the presence of the female had any direct impact on the discussion or the decision cannot be evaluated. Yet Shana Stone admits "this place has really taken its toll. There is no sanity in seeking a position at one of the nation's most competitive institutions unless you are willing to fight every step of the way."

Movement toward parity in salaries has been substantial in the past two decades, partly because of continuing pressure by women professionals to make it happen. Institutions where disparities still exist are fewer in number; but outmoded attitudes and practices often have lingering deaths.

Some suggested actions that women scientists can take to assist in the demise of salary differentials include:

- Encouraging public disclosure by the institution of its salary scales and ranges at each professional grade level

- Encouraging legislation and union agreements guaranteeing equal treatment of females and males in promotion, salary, and tenure decisions

- Assisting in legal actions and supporting administrative policies that assure parity in salaries of female and male cohorts

- Participating in local, regional, national, and international organizations of women scientists and supporting agenda items that lead to scrutiny of and publicity about institutional salary practices

- Assuring maximum representation of women on faculty senates, faculty policy committees, and counterpart groups

- Ensuring that, because of their numerical minority, women faculty members are not overly encumbered with nonessential committee and other assignments that detract from their scientific productivity, hence interfering with promotions

- Adopting more assertive tactics when individual promotion and salary discussions are held with department heads, deans, and other administrators—including, if necessary, legal steps in instances where they seem warranted.

The route to a favorable tenure decision is of course intimately interwoven with individual promotion and salary history and with a record of excellence as a scientist. Accepting the reality that, at least for the immediate future, tenure decisions will be made by groups consisting mostly of men, and approved or disapproved by male administrators, it becomes important to minimize career events that can be used as negative elements in the decision-making process. Included here would be such factors as:

- Time spent in nontenure track positions (such as research associateships)

- Time spent in leave for childbirth or child care (this can be especially true in institutions that do not have a stated "stop-the-clock" policy)

- Requests for less than full-time employment, which, regardless of the justification, can be interpreted as an indication of lack of professional commitment

- Restricted office and classroom hours, possibly necessitated by lack of adequate child care or other family support facilities

- Bringing a child or children into the work environment, even for limited hours

- Significant gaps, required for personal reasons, in the continuity of job histories.

These elements can impinge directly on tenure decisions affecting women professionals since their societal roles as wives and mothers may be viewed as taking precedence over their roles as professionals.

In response to our persistent probing, a number of outstanding women scientists have tried to encapsulate, from their own experience, some universal "rules of the road" to achieving success. Two of those capsules are listed here.

One woman scientist expressed her personal hang-up about the markedly sloppy and occasionally bizarre clothes worn on the job by some women professionals. She urged a self-imposed dress code, possibly leaning toward the conservative, pointing out the reality that in the academic environment it is wiser for women to dress in ways that will not detract from their work. After all, their male colleagues will be participating in decisions about career advances. (This woman scientist with the dress code complex had nothing to say about the apparent freedom that some male scientists have to wear any poorly fitting, out-of-date, sloppy outfit without fear of career consequences.)

From a productive woman scientist who also serves as an academic dean: "You must be bright, enthusiastic, and productive. You must be ready to exploit all legitimate angles. You must believe in yourself and maintain a consistently high profile."

SUMMARY

Primary career goals of women scientists can be identified as: the contribution of significant new scientific/technical information; acquisition of respect and recognition from peers; and acceptance as an equal participant by the "system" of science. Problems impeding achievement of career goals seem to center on institutional policies and practices—salaries, promotions, and tenure—even though significant progress has been made during the past two decades.

- Gender parity in salaries has been achieved in many academic, research, and industrial organizations, but in others, women scientists are still paid less at a given professional grade level than are men. The differential in the latter institutions has narrowed in the past decade, but still averages as much as 10 percent overall.

- Comparable gender parity in rates of promotion and acquisition of tenure has not yet been achieved; women scientists are still aggregated in the lower ranks, often in nontenured positions.

- Gender intrusions in tenure decisions are difficult to evaluate, but the percentages of tenured or tenure track positions occupied by males are disproportionately higher.

Despite such continuing gender-related inequities at the institutional level, only scattered remains of bias exist in the larger community of science. Professional recognition, in the form of awards and election to society offices, has become remarkably gender-free and based principally on merit and performance.

One persistent impediment to equal professional recognition is the puzzling difference in scientific productivity (as measured by publications) between male and female scientists. A number of plausible explanations for the disparity— motivation, demands of childbearing and rearing, differences in early acculturation—have been proposed. It is apparent, though, that those women who publish extensively have a better opportunity for senior positions at major universities than was true in the past.

REFERENCES

1. Vivian Gornick, *Women in Science: Portraits from a World in Transition* (Simon and Schuster, New York, 1983), p. 145.
2. Betty L. Harragan, *Games Mother Never Taught You* (Warner Books, New York, 1977), p. 1–399.
3. John T. Bruer, Women in science: Lack of full participation, *Science* 221(4618), 2312 (30 September 1983).
4. Cathy Shaw, Women still lag in faculty pay, union study says. *Miami Herald* (1 January 1989), p. C-1.
5. Jonathan R. Cole and Harriet Zuckerman, Marriage, motherhood and research performance in science, *Scientific American* 255(2), 119 (1987).
6. Jonathan R. Cole, *Fair Science: Women in the Scientific Community* (Free Press, New York, 1979).

7. George E. Johnson and Frank P. Stafford, Pecuniary awards to men and women faculty, in *Academic Rewards in Higher Education*, eds. Darrell R. Lewis and William E. Beeker, Jr. (Ballinger, Cambridge, MA, 1979), pp. 231–243.

8. National Education Association, *Salaries Paid and Salary-Related Practices in Higher Education* (National Education Association, Washington, D.C., 1972), Research Report 1972-R5).

9. Alan E. Bayer and Helen S. Astin, Sex differentials in the academic reward system, *Science* 188(1421), 796 (23 May 1975).

10. Stanley Budner and John Meyer, Women professors, cited in Jessie Bernard *Academic Women*, (Pennsylvania State University Press, University Park, PA, 1964).

11. Jonathan R. Cole and Stephen Cole, *Social Stratification in Science* (University of Chicago Press, Chicago, 1973), p. 134–139.

12. Jessie Bernard, Benchmark for the 80's, in *Handbook for Women Scholars*, eds. Mary L. Spencer, Monika Kehoe, and Karen Speece (Americans Behavioral Research Corporation, San Francisco, 1982), p. 78.

CHAPTER 3

EARLY EDUCATION AND TRAINING
OF WOMEN SCIENTISTS

Early childhood sense of wonder; keeping curiosity alive;
fostering obsession; pump versus filter.

Scientists have been depicted as "merely children who have not grown up—curious about the natural world and repeatedly asking 'Why?'" If this description is acceptable, then the roots of becoming a scientist begin at a very early age. Infants and toddlers are overcome with a sense of wonder.[1] Unfortunately, this sense of wonder fades for most students, but seemingly to a greater degree for females than for males.

News stories from the late 1980s and early 1990s brought repeated accusations of scientific and math illiteracy in American schoolchildren in grades K–12 as well as among college students. As an example, when American high school students were compared with counterparts in twelve other countries, they ranked *eleventh* in chemistry achievement tests, *ninth* in physics, and *last* in biology (the most popular science course taken by American students).[2] These dismal findings must be viewed against the background that the adults tested also acquired F's in science. A 1988 National Science Foundation survey revealed that just 45 percent of adults knew that the earth orbits the sun in one year; only 37 percent recalled that dinosaurs lived before the earliest humans; 43 percent recalled correctly that electrons are smaller than atoms; and 36 percent responded that lasers do not use sound waves.[3]

EARLY EDUCATION

A reconstruction of science and mathematics education in American schools is beginning, with major initiatives being undertaken by the American Association for the Advancement of Science (AAAS), the National Academy of Science (NAS), the National Council of Teachers of Mathematics (NCTM), the National Science Foundation (NSF), and the Department of Education.

The dilemma? S. Begley and associates[3] put it this way: "Educators take children who demand 'why?' and 'how?', who poke and drop and squeeze like the most exuberant experimenters, and turn them off to science completely and irreversibly." Those authors then went on to quote the Nobel prize-winning physicist from the University of Chicago, Dr. Leon Lederman: "Schools take naturally curious, natural scientists and manage to beat the curiosity right out of them." The report indicated that by third grade 50 percent of all American students claim to dislike science; and by grade eight 80 percent of them stated that they disliked science.

The present challenge is to educate and inspire both male and female students. The remedy will be reconstruction of science and mathematics programs, commencing with retraining and enrichment of teachers and faculty. Several recommendations have been presented in AAAS and NAS reports. Although reliable statistics are difficult to obtain, most mathematicians, scientists, and educators feel that the filtering out of females is more extreme and deserves special attention. The challenge and opportunity will be to construct programs that are free from gender bias.

There is a growing concern that the so-called gender gap in science, math, and engineering professionals is due to more than a lack of stimulation of young women, but is in fact the result of a "turnoff" of young women. There are a number of common perceptions about science and math education, some of which we agree with and some of which we disagree with. We will explore in some detail the credibility and accuracy of these perceptions:

- There are basic biological differences between males and females that contribute to the gender gap in science and math.

- Experiences contributing to motivation for a career in science commence as early as junior high school.

- Females are less interested than males in K–12 science and math.

- Females are less adept than males in science and math subjects.

- Traditional science and math classroom and laboratory activities tend to favor males.

- Females are less adept than males in dealing with the normal give-and-take of team projects.

- Females aspiring to be scientists are stymied by an absence of role models and mentors.

Biological differences? Indeed. Yet we have no clear explanation of why or how these differences should affect learning, aspiration, and motivation. Biological differences between males and females are numerous: physical anatomy and body size differ; hormonal composition differs with a maximum difference existing over the peak reproductive years; chromosomal complement differs; and recent evidence indicates that certain aspects of the brain are different in males and in females.

Anatomy and hormonal differences become increasingly obvious as a child reaches puberty, which normally occurs in junior high school/middle school. As one Ph.D. scientist who is presently a junior high school science teacher put it:

> It is extremely difficult to override the intrinsic physical and chemical cues in teenagers. If I can help to (1) keep the students curious, (2) increase self-understanding, and (3) develop self-esteem, then I feel that I have done some good. Maturation of a scientific mind comes somewhat later. . . . BUT YOU HAVE GOT TO KEEP THE CURIOSITY AND OBSESSION ALIVE.

What has come as somewhat of a surprise is a phenomenon present throughout development, from a young child to old age. A part of the brain seems to do one thing in men and something else in women. Dr. Sandra Witelson, of McMaster University in Toronto, studied the fibers that connect the right and left hemispheres in the brain, the corpus callosum.[4] In this study, Dr. Witelson found that the isthmus section was larger in women than in men. Examination of the isthmus size can be correlated to left-handedness in men but not so in women. Even when such differences are well documented, it is difficult to determine what these differences mean. Rockefeller University's Dr. Bruce McEwan suspects that there are consistent differences between male and female brains but acknowledges that it has not been possible to distinguish which are genetic and which are environmental. The structure of the brain is altered by experience,[5] and the experiences of males and females are obviously quite different.

Brain differences in males and females will get intense study in the years to come. Some differences are possibly based on the chromosomal difference of females having a double complement of X chromosome while the male has an X and a Y chromosome. Mapping of both the X and Y chromosome will permit knowledge of which proteins are encoded by these regions of DNA, and subsequently what functions might be altered by the presence or absence of one or the other.

The perceptions that females are less interested than males in science and math, are less adept than males in science and math, and that traditional classrooms and laboratories as well as standardized tests tend to favor males cannot be refuted with ease. What we can see is that both males and females fall into one of several learning style categories—dynamic, innovative, common sense, and analytical learners.[6] Our scientist respondents also seemed to fall into one or another category, indicating that there is no single type of learning style that produces excellent scientists (Figure 5).

Other detailed studies have been reported. An examination of results of standardized tests disclosed that over

Dynamic Learners	Innovative Learners
Seek hidden possibilities.	Seek meaning.
Need to know what can be done with things.	Need to be involved personally.
Learn by trial and error, self-discovery.	Learn by listening and sharing ideas.
Touch reality.	Absorb reality.
Perceive information concretely and process it actively.	Perceive information concretely and process it reflectively.
Adaptable to change and relish it; like variety and excel in situations calling for flexibility.	Interested in people and culture. They are divergent thinkers who believe in their own experience, excel in viewing concrete situations from many perspectives, and model themselves on those they respect.
Tend to take risks, at ease with people but sometimes seen as pushy. Often reach accurate conclusions in the absence of logical justification.	
Function by acting and testing experience.	Function through social interaction.
Strength: Action, carrying out plans.	*Strength:* Innovation and imagination. They are idea people.
Goals: To make things happen, to bring action to concepts.	*Goals:* Self-involvement in important issues, bringing unity to diversity.
Favorite question: "What can this become?"	*Favorite questions:* "Why or why not?"
Careers: Marketing, sales, action-oriented managerial jobs.	*Careers:* Counseling, personnel, humanities, organizational development.

FIGURE 5. Learning Style Type Grid. SOURCE: David A. Kolb.[6]

Common Sense Learners	Analytic Learners
Seek usability.	Seek facts.
Need to know how things work.	Need to know what the experts think.
Learn by testing theories in ways that seem sensible.	Learn by thinking through ideas. They form reality.
They edit reality.	
Perceive information abstractly and process it actively.	Perceive information abstractly and process it reflectively.
Use factual data to build designed concepts; need hands-on experiences, enjoy solving problems, resent being given answers, restrict judgment to concrete things, have limited tolerance for "fuzzy" ideas. They need to know how things they are asked to do will help in "real life."	Less interested in people than ideas and concepts; they critique information and are data collectors. Thorough and industrious, they will reexamine facts if situations perplex them.
	They enjoy traditional classrooms.
	Schools are designed for these learners.
Function through inferences drawn from sensory experience.	Function by adapting to experts.
Strength: Practical application of ideas.	*Strength:* Creating concepts and models.
Goal: To bring their view of present into line with future security.	*Goals:* Self-satisfaction and intellectual recognition.
Favorite question: "How does this work?"	*Favorite question:* "What?"
Careers: Engineering, physical sciences, nursing, technicians.	*Careers:* Basic sciences, math, research, planning departments.

FIGURE 5. (*Continued*)

the past twenty years, men's scores were higher than
women's scores for standardized tests such as the Scholastic
Aptitude Test (SAT) and the American College Testing Program (ACT). Test results from the SAT are given in Figure 6.
In articles from the *New York Times*,[7,8] discussing this gender
gap in test results, the cause was found to be elusive.
"Increasingly, the explanations for gender distinctions focus
on one phenomenon—sometime during the junior high
school years, at an age when children begin to cement their
sexual identity, many girls give up on math and science."
One article goes on to say that

> Critics contend that many tests are biased, containing questions
> on topics like sports that are congenial to men. But the Educa-

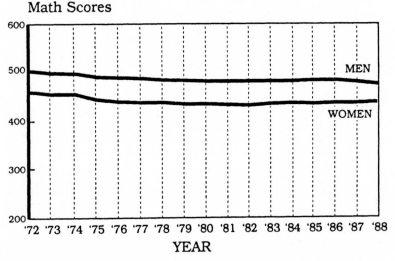

FIGURE 6. Scholastic Aptitude Test (SAT) scores from 1972 to 1988. Mean
of male subpopulation (top) and mean of female subpopulation (bottom).
SOURCE: Laura Mansnerus.[7]

tional Testing Service which writes the SAT, the most commonly taken college admissions exam, has long used rigorous procedures that it contends screen out questions in which men outperform women of similar ability.[7]

Other studies suggest that females are more thorough and therefore take more time to complete and therefore do poorly on *timed* tests like the SAT and ACT. A 1991 report indicates that SAT scores fell to an eight-year low, with males averaging 923 and females 871. "Most of the difference comes in a 44 point difference in math."[9]

Another phenomenon that appears to be real is that there are additional home pressures cast on young girls in the increasing numbers of single-parent households. The phenomenon may be present to a lesser extent in two-parent households. Frequently the oldest daughter, not the oldest child, is given a series of home chores to complete in the parent's absent hours. If there are younger siblings, babysitting is expected. The freedom to participate in pleasurable organized after-school science and nature clubs is lost. Lost too is the simple opportunity to linger at creekside on the route home. Many teachers report this as a concern of which educators must become aware. Whether or not this contributes to the gender gap is unknown but worthy of serious study. It is a bit ironic that in households with employed women the young girls of the household are still cast in traditional roles.

A related issue concerns the gender gap in interpersonal relations. In question here is the business of doing science, not concept perception and processing. Females often seem less adept than males in dealing with the normal give-and-take of team projects, even though women are noted for their social skills generally. Some studies have suggested that the childhood activities of males and females are far different. Boys are often given science or technical objects as gifts, whereas girls get dolls or clothing. The norm of females playing noncompetitive games (jacks, dolls, jump rope) and males engaging in competitive sports (soccer, football, baseball, basketball, hockey) is changing. "Some sports previously considered male-oriented have caught on with women."[10] Yet there is now a new trend. Women are becoming actively engaged in competitive team sports—but the teams end up being single-gender. The outcome may be different but equally challenging. To develop the desired outcome of team study and group learning, men and women, blacks and whites must remain integrated in all phases of activity.

The perception that experiences contributing to motivation for a career in science commence as early as junior high school is debatable. When investigators first looked into the question of the age at which a student is molded into becoming a career scientist, "college" was a typical reply. Yet when we questioned our respondents, inevitably the response was that the roots of their curiosity were well established in elementary school. Often the scenario included one special teacher who was dedicated to keeping curiosity alive, to fostering obsession, to encouraging experimentation. Many respondents indicated that they kept an active relationship with that teacher throughout their student days. Many indicated that they continued an active relationship as adults. This suggests that role models and mentors are not always other career scientists, but are also educators trained in science who have dedicated their working lives to teaching.

One of the highest-ranking people in the Office of Budget and Control at the National Science Foundation— Sandra D. Toye—addressed the Women's Chemist Committee in the autumn of 1990 and presented the NSF view of "Women in Science and Engineering."[11] She claimed that science has an exclusionary bias that rejects the majority of students, with women being rejected to a greater degree than men. As reported in the Women's Chemists Newsletter: "Those who would get into science must overcome hurdle after hurdle, as the process 'weeds out' rather than tries to 'bring in.' . . . A change in focus is needed to stop 'screening out' and to start 'screening in' beginning with the early undergraduate programs."

The mathematicians have stated the concept in this way: What is needed is a *pump* system in education versus a *filter* system.[12] Special target programs are emerging, including PACTS (Parents and Children for Terrific Science), a small grant program by the American Chemical Society for young girls, and the WEEA Program (Department of Education Women's Educational Equity Act). These programs are expanding. Specific information can be found in the Appendix of this book.

Females aspiring to be scientists are also stymied by the scarcity of role models and mentors. The number of role models is growing. To gain perspective on this growth, we have plotted the percentage of women scientists in the total scientific work force against a background of significant dates in science and discovery for two millennia, and for the twentieth century (Figure 7). A role model is still not always within reach, but many role model programs are being initiated: for example, the *Science-by-Mail*™ program sponsored by the Museum of Science in Boston for students in grades 4–9.

But there is a downside to this subject that receives too little attention—that of the female scientist as a *negative* role model. Students are not usually dummies: they see a

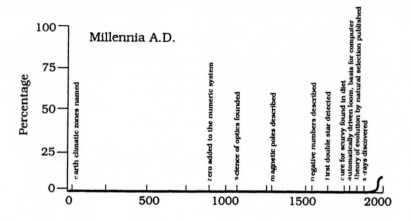

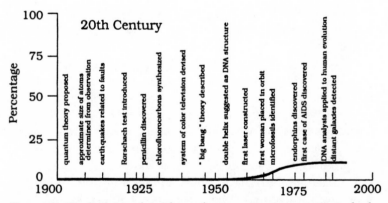

FIGURE 7. Gradual rise of numbers of women in science against a background of sampling of significant dates of science and discovery, presented as millennia (top) and 20th century (bottom).

continuously overstressed woman scientist trying to balance career and household demands, but being judged by the same professional criteria that are applied to men—some of whom seem to have enormous amounts of free time. Their reaction may well be, "Who needs this?" or "Where is all the touted joy and satisfaction in this kind of existence?" Their pragmatic assessment might be, "Maybe I don't want to be a scientist after all."

A recent study by Jones of 1300 students in 60 physical science and chemistry classes in North Carolina, in Sandy Hill's article "Researchers Find Subtle Biases Favor Boys in Science Classes,"[13] confirms that teachers do respond differently to male and female students. In general, teachers call on girls less often, encourage girls less, and tease girls more often than boys. Posters in the classroom and photographs in textbooks still convey the impression that the brain of a scientist or mathematician is part of the male anatomy.

Teachers are usually unaware of their biases, but can be made sensitive to their actions by such simple steps as videotaping their classroom activities or requesting an external observer to oversee a class or two. When teachers initiate these actions, the results are more noteworthy than if such actions are taken because of an administrative dictate.

An important point is that the responsibility is not solely with the teachers. Parents have even greater influence and must provide suitable women role models, books, chemistry sets, rocket kits, microscopes, hands-on experiences, and continual "go for the gold" encouragement, and they can influence the children's school administrators and teachers to do the same. If parents, teachers, and administrators all hold a uniform vision, then equity can become a reality.

Even in our encouragement are hidden messages. "You

can do it." "You can become part of an exciting team." "You can be a valued team member." "You can make a contribution." The result of these words of encouragement is what might be called the bronze complex. One becomes satisfied when aspirations are to join the team, participate, be a recognized player. But in reality, each of these statements of encouragement is a minimal statement. Instead, the encouragement must be strengthened with advice such as, "You can come up with the best idea in the class," "You can be the leader of the team," "You can become the best."

SUGGESTED ACTIONS

Our suggested actions have been adapted from Hill[13]:

- Provide posters for the classroom that show 50:50 male:female participation in the processes of science, or if you need to neutralize an existing bias, provide the female complement showing women in nontraditional careers, in leadership roles.

- Seek out and provide names and contact information (affiliation, education and training, address, phone, fax, electronic mail) of women and minorities who would be willing to speak to a science and/or math class, or at career day, or answer questions about their specialty via electronic mail. Volunteer yourself.

- Offer your assistance to the classroom teacher by leading a small group in class. The rationale is that girls and minorities learn scientific principles easier in small groups where discussion and cooperative exploration are facilitated.

- Suggest and organize a parents' night where parents and students can learn of new nontraditional careers for women and minorities.

- Work with school administrators and teachers to assure that female and male students get equal access and encouragement to use laboratory, shop, and computer equipment both during classroom hours and after-class hours.

- Help with science/math clubs and competitions, with special attention and encouragement to females. Work to reduce feelings of alienation. Remember, the fewer the females, the greater the support females need. Once numbers approach 50:50, a critical mass is reached that removes feelings of anxiety, isolation, and alienation.

- Encourage women math and science speakers capable of relating to the layperson to address service clubs such as Kiwanis, Lions, and Rotary. The public attitude needs a change, too.

- When you encourage young women, don't merely praise them for being participants in the scientific enterprise. Praise them, yes, but then convince them that they can be leaders—the best—to go for the gold.

- Organize regional programs of selected students that are 50:50, with special attention to freedom from gender bias in the presentation, content, and in all written and spoken material. This provides female students with opportunities to relate to one another outside of their institutions. Encourage pen-pal relationships. This practice, early on, assures young women that they are not alone in their interests and intellectual drive.

• Organize vacation and summer science/math/computer camps that are 50:50 female:male. This can be a healthy way to provide a positive peer group of curious, striving students of similar interests. These arrangements help students to calibrate themselves with others from other schools.

SUMMARY

Those who care about the future of women in science acknowledge that advances are being made, but are often frustrated at the slow rate of progress. Outdated attitudes are slow to die. Accordingly, invisible barriers still remain that stunt females' motivation and aspirations in science. Yet, it is acknowledged by the major funding agencies, NSF in particular, that women and minorities will be needed to supply scientists, technologists, and engineers for the future. These people must be suitably educated to find desirable positions in institutions where the climate in the workplace is comfortable. The situation is not merely that women are capable of doing good science or great science and therefore should be permitted entry into scientific and engineering professions. Equity must be so commonplace that it becomes a nonissue. Special efforts must be made to *attract* women, and once they are at an institution, the climate there must be so congenial and supportive that women will want to stay. The desire is for *absence* of constraints.

In reality, the issues go beyond caring about the individual woman scientist or even about women scientists collectively. The situation is assessed by some that women will be the key to saving science. Mary Clutter, Assistant Director of the National Science Foundation, states: "Those who truly care about the future of science know that women represent an untapped resource."[14] There is a long way to go. While women constitute nearly 50 percent of the total professional work force in the United States, only 15 percent of working scientists are women.

How can the climate be improved for women scientists? The responses and opinions are multiple and varied. Many experimental programs are being tested. Within the National Science Foundation there are innovative efforts to encourage women. These programs, as described in summary information provided by NSF, include: Faculty Awards for Women (FAW), to recognize and retain the nation's most outstanding and promising women scientists and engineers in academic careers of research and teaching; The Research Initiation Considerations, which are onetime awards to provide opportunities for women scientists or engineers who have not served as principal investigators or co-principal investigators on individual federal research awards; Research Planning Grants, which are onetime awards to provide opportunities for women scientists or engineers who have not served as principal investigators or co-principal investigators on individual federal research awards, or whose careers have been interrupted for at least two of the past five years; and Career Advancement Awards, which are awards to expand and advance the applicant's research career. These awards are available to all women scientists and engineers, but especially appropriate for junior women faculty is the Visiting Professorships for Women (VPW) program, which funds experienced women scientists and engineers to serve as visiting faculty members at host institutions. In addition to undertaking advanced research at a university/research institution, the visiting professor undertakes lecturing and other interactions to increase the visibility of women scientists in the academic environment. Specific contact information for these programs is provided in the Appendix.

There are intriguing examples of K–12 educational reform, where teachers undergo sensitivity training and thus try to make eye contact and call on girls as frequently as on boys for responses to questions. Experimental schools already exist that will permit testing of near-equity. One example is the North Carolina School for Science and Mathematics in Durham. This model school was established in 1980 to provide access to research and knowledge of science for high school juniors and seniors. The enrollment of 550 students in residence is based on a highly selective process and has, by design, 50 percent females and 50 percent males. Eleven states have started comparable schools modeled after North Carolina. Students from the early years of the North Carolina program have now graduated from college or university and from graduate or medical schools and are part of the work force. Tracking women within this sample will be highly informative. Specific questions of interest are:

1. What percentage of females entered graduate school compared with females from normal high school populations? Compared with females from co-ed prep school populations? Compared with females from female prep school populations?

2. Within the North Carolina School for Science and Mathematics, how did females compare with males

on test scores such as the Scholastic Aptitude Test? the Graduate Record Exam?

3. Is there evidence that the females in this subpopulation are being filtered out or pumped in to the mainstream of preparation for careers in science?

REFERENCES

1. Rachel Carson, *A Sense of Wonder* (Perennial Library, New York, 1956, 1984).
2. G. Crowley, K. Springen, T. Barrett, and M. Hager, Not just for nerds, *Newsweek* (9 April 1990), pp. 52–54.
3. S. Begley, K. Springen, M. Hager, T. Barrett, and N. Joseph, Rx for learning: There's no secret about how to teach science, *Newsweek* (9 April 1990), pp. 55–64.
4. Patricia Keegan, Playing favorites, *New York Times Special Section: The Gender Card* (September 1989), p. Educ. 26.
5. Gina Kolata, Mind blowing? *New York Times Special Section: The Gender Card* (September 1989), p. Educ. 15.
6. David A. Kolb, Learning Style Type Grid (Excel, Inc., Oakbrook, IL, 1976).
7. Laura Mansnerus, The S.A.T. puzzle, *New York Times Special Section: The Gender Card* (September 1989), pp. Educ. 23–24.
8. Joseph Berger, All in the game, *New York Times Special Section: The Gender Card* (September 1989), pp. Educ. 23.
9. Pat Ordovensky, S.A.T. scores fall to 8-year low, *USA Today* (27 August 1991), p. 1A, 1D, 8D.
10. Elaine Louie, Unequal contest, *New York Times Special Section: The Gender Card* (September 1989), pp. Educ. 28–29.
11. Sandra D. Toye, Women in science and engineering, Address to Women's Chemist Committee, in *Women Chemists* (American Chemical Society, Washington, D.C., 1990), pp. 1, 2.
12. Billy Goodman, Toward a pump, not a filter, *Mosaic* 22(2), 12 (1991).
13. Sandy Hill, Researchers find subtle biases favor boys in science classes, *The Boston Sunday Globe* (21 July 1991), p. 81.
14. Robin Eisner, Science's future: Do women hold the key? *The Scientist* (15 October 1990), 4(20), 1.

CHAPTER 4

A TIME LINE APPROACH
TO WOMEN'S LIFE-STYLES
AS CAREER SCIENTISTS

*Multiple choice; managing multiple priorities; balance;
selecting an image; projecting that image; coping with
devaluation; guilt control; the C+ syndrome.*

We asked some of our more senior male respondents, "Has
your attitude toward women in science changed, and if so,
when? From what to what?" The answers given by these
men were remarkably similar. The breakthrough seemed to
be in the early 1970s. On further prodding, they claimed that
the change came about when females recognized that they
could be "women and scientists" simultaneously—when the
pressures slackened to mimic males in order to break into
the still-predominantly male world of science. Gradually
women in science were no longer looked upon as "husband
hunting" and were taken seriously because they would

continue with their careers "even" when they married and/
or had children.

TAKING WOMEN SERIOUSLY

Granted, some of our male respondents said that their
attitudes had not changed markedly, but it is encouraging to
find greater acceptance of women scientists by so many
"reconstructed" males. Many perceptions persist about the
life-styles of females in science. These include:

- Women scientists are more concerned than men scientists about the quality of life in the work environment.

- Women scientists are more concerned than men scientists about the balance between personal and professional lives.

- Women who combine science, marriage, and motherhood are less productive than unmarried women
  scientists.

- Women who combine science, marriage, and motherhood must master guilt.

- It is abnormal for women to pursue their own dreams
  rather than to be an adjunct to a man's dream.

- With career couples, the wife is usually the "trailing"
  spouse.

- The aim of women's liberation is to reduce the price
  for fulfillment and thereby make it within reach for all
  women.

We have set aside a later chapter (Chapter 13) to discuss the career stages of women scientists, but here we intend to discuss various aspects of the life-styles of women scientists, and how the choice of life-style does or does not impact on productivity as a scientist. To do this we will need to define parts of the personality with which we are dealing. We will adopt a general sociological description of an individual that includes three aspects. These are: the intellectual self, the "caring" emotional self, and the sensual/sexual self. Each of these aspects is considered highly developed in a "normal, mature" adult. Each aspect requires learning and skill.

Until the last few decades, males were expected to indulge in intellectual and sensual/sexual development. Women were encouraged to indulge in emotional and sensual/sexual development. (We are reminded of the trite phrase "the man is the head of the family but the woman is the heart of the family.") The feminist movement ground swell in the 1960s started to challenge these precepts, and little by little, the financial responsibilities plus household and child-rearing responsibilities became shared. The result is that a 1990s individual, male or female, is more often than not quite well developed in all aspects—intellectual, emotional, and sensual/sexual.

The women scientists we interviewed reported that they felt that they were more concerned about a balance in their personal and professional lives than the men that they worked with or the men that they were married to. Unmarried females considered balance as important as did married females. Decisions in female scientists' lives are often determined by numerous considerations, not merely potential for intellectual development. Paula K. Burkhard, Acting Director of the Office of Research and Sponsored Programs at the

University of Oregon (an institution that is highly innovative in equal employment opportunities—so much so that female employees consider gender a "nonissue"), commented that "Women in academia tend to make their acceptance decisions on more holistic reasons [than men do]."[1] The outcry for maternity and child care options and "some mechanism to handle everyday affairs" is not new. In 1885, mathematician Sofia Kovalevskaia wrote to a friend, "All these stupid but unpostponable everyday affairs are a serious test of my patience, and I begin to understand why men treasure good, practical housewives so highly. Were I a man, I'd choose myself a beautiful little housewife who'd free me from all this."[2]

What then is the life-style of a reconstructed woman scientist some 100+ years later? Although women have been liberated to the extent that they pursue their own vision and dreams independently of the dreams of their spouse, it is rare that the caring/emotional aspect of their personality takes second priority. Most married women scientists still insist that their children and family have first priority. Recognizing this, in Figure 8 we have attempted to portray the lives of successful women scientists along a time line of years from birth (at left of diagram) until the present (at right of diagram). Obviously, older women will have a longer time line. The time when degrees were awarded are indicated by symbols along the time line. Superimposed on this time line are noncareer priority events and responsibilities. When married or coupled with a significant other, an angular line is drawn to represent a partner, and that line persists until divorce or separation (denoted by a break in the line) and/or death (denoted by a vertical mark at the end of the line). Offspring are noted, again, with pregnancy as a deeper angular line that persists for the life of the individual. Male offspring are placed below the main time line and female offspring are placed above.

Figure 8 is an example of several time lines for women who have been judged by their peers as highly successful, influential, and accomplished scientists (based on individual identity, research publications, funded competitive proposals, and job placement). No clear pattern of life-style emerges. Yet the responses of these women to our questions about "How would you define success in science?" "What are the things about doing science that you find to be most satisfying?" and "Are you satisfied with your rate of career progression?" were highly uniform. The point here is that while the choices and expectation in personal life-style are quite different, their expectations and performances in careers are remarkably similar.

Each of these women (whose time line is represented in the diagrams) also admitted that they were victims of the so-called "C+ syndrome." Each woman scientist had been a good student in high school and college and graduate school, with A's (sometimes B+'s) in everything. Now, with multiple priorities, they were feeling like they were achieving only a C+ in everything: professional growth, marriage, motherhood, and community service.

For women, there are multiple choices, many of which deviate from the traditional or societal norms. First, a female must define her own goal, her vision, expressing her career dreams/research obsession. Next she settles on a life with or without a partner. If she selects a partner, then the partner must be respectful of her dreams. The next major step is obtaining a position that is consistent with her dreams. Another decision is whether or not to have children. If yes, then, how many? When? How will child rearing be accomplished? Fulfillment and maximum productivity will only be

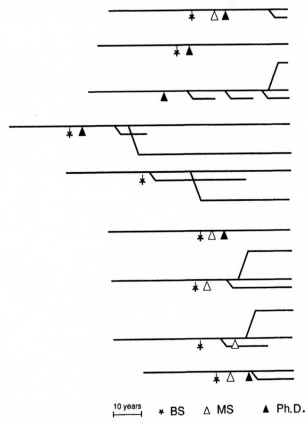

FIGURE 8. Time line of various successful scientists. The scale bar is 10 years. The academic degrees are symbolized as designated, * = bachelor's degree, Δ = master's degree, and ▲ = Ph.D. Spouses are designated by angular line and lifeline beneath the main line. Children are designated by angular line with females above the woman scientist's lifeline and males below the woman scientist's lifeline.

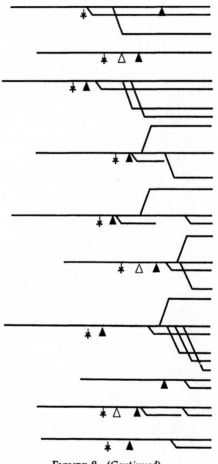

FIGURE 8. (*Continued*)

achieved once she has found *comfort* and *security* in her home
with appropriate privacy, affection, and intimacy, and *comfort* in her job, with appropriate recognition, promotion, and
rewards.

The priorities for women rapidly become multiples, and
one individual cannot do it all, so interdependency results,
even for the most independent women. Comfort is only
reached once a solid support network is built. The resulting
dependency by so-called independent women has two notable negative facets: guilt and loss of spontaneity. Controlling
guilt is an agony common to life-styles of career women.
Regaining spontaneity and managing multiple priorities are
challenges. How are they accomplished?

We share here the reflections of one of our respondents,
Dr. Claudine Mercer, a microbiologist.

> My realization came when I was talking with my husband,
> also a career scientist, and he stated that he never did anything
> that he didn't want to do! I found this quite stunning as I felt that
> I spent a lion's share of each day cycling through the necessities
> so that I could get to the parts of my responsibilities which I
> value and enjoy most. This was true for both science and home.
> I found this difficult to reconcile. At that point, I started to
> analyze just how I operate. It struck me that I had a computer-
> like main menu with choices of other menus, each mutually
> exclusive. Indeed, when I awake in the morning, I decide I want
> to take the extra time today to be warm and friendly, and in
> doing so, my productivity drops markedly. If I need a highly
> productive day, seeking to finalize a proposal or manuscript, I
> wake and decide, 'All systems go on the document today.' If
> the day is destined for laboratory experiments, I wake and
> prepare for the unexpected, both in terms of science and for
> calling upon my support network for alternatives for the children
> in the probable case that the experiment runs late into the day or
> evening.
> For me, parenting priorities come before professional priori-

ties, so more realistically it is all systems go on the proposal *after* I have gotten the kids organized for preschool, packed their lunches, etc. Even though my husband and I try to share the day-to-day *predictable* responsibilities and chores, I still take care of most of the day-to-day problems.

These are the words of one of our respondents, but we have noted similar reflections by others. The behavior style described by Dr. Mercer might be termed calculated behavior. Because it is so general for women, we have presented it in Figure 9. The male counterpart (offered by one male respondent) is given in Figure 10. At every junction, a decision is made that opens the choices for the next mode of operation. The couple in the illustration in fact attempt to share household chores and parenting quite equally. Why then does the female feel overwhelmed while the male can feel completely at ease? Although they are experts and have separate identities in different subfields, both husband and wife work in the same university environment, the same department, have the same bosses, and are subject to the same institutional policies. They live in the same home, have the same children, and share family challenges.

We went on to explore this phenomenon in depth with others we interviewed. Indeed, many females claim that the only way in which they can be efficient enough to survive is to run through a computerlike menu, spending as little time as necessary on the items before them. Only when they have done enough in the categories of parenting and partnering to free them from extreme guilt feelings, do they become absorbed in being the science professional. This is a strong argument in favor of institutional day-care programs. This transition is of course done on a daily basis, but the energy up front is sizable and can be exhausting. (Yet once a system has been set up, be it babysitter, preschool, or camp, there is

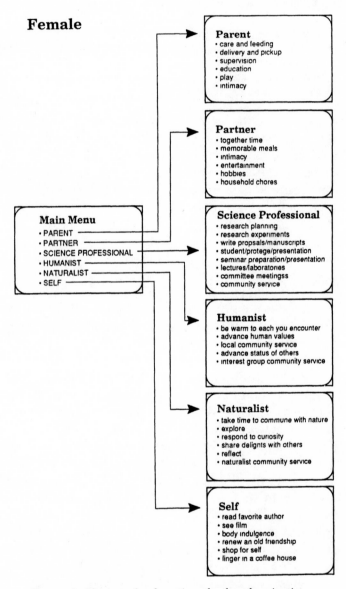

FIGURE 9. Menu and subroutines for female scientists.

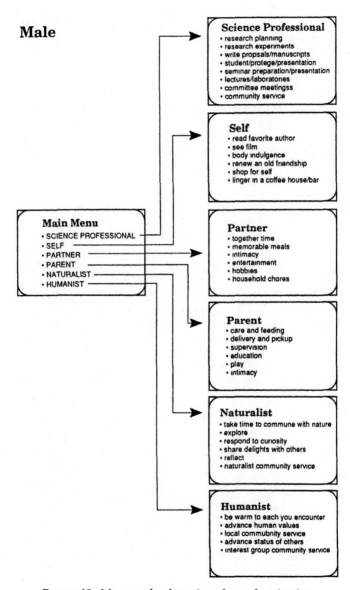

FIGURE 10. Menu and subroutines for male scientists.

no great stress.) Getting to the subroutines of the scientific professional must occur daily; the humanist and naturalist are likely to occur only once per week or once per month. Getting to the subroutine of self is often overlooked for months to years by women scientists. There are few guilt feelings that accompany self-deprivation. Women are susceptible to burnout at ages far younger than their male counterparts.

> A nontechnical analysis of Dr. Joe Mercer, Claudine's husband, indicates that he always watches his football games, heads to the pub with his laboratory buddies after work, exercises daily (weather permitting) in his rowing scull, and considers being a science professional as his number-one priority. He takes pride in his care giving to his children. At each meal he is involved in their care and feeding; he delivers the children to preschool; he supervises the children on the home front; he reads to/with them; he takes them to the YMCA for swimming lessons; he plays active games with them; he takes time to be their "best friend" and is a good listener. Additionally, he is a good role model for his young "new-age" sons, as he is often pushing the vacuum cleaner; he is often putting laundry from the washer into the dryer or out on the clothes line; he is often doing the dishes, making supper, or marketing.

One wonders, yes, but who decides *when* and *what* jobs to do? Who *plans* and sets the responsibilities? If the husband doesn't do some of the day-to-day chores, who picks up the slack?

The ethos is changing. Married males are responding to the challenge of joint parenting and joint performing of household chores on a daily basis. Males involved in the care of home and family often find little time for self-indulgence, but they do find the time. What does seem to be a difference between males and females is the guilt quotient. Women feel some degree of guilt when they do not do something; males do not seem to suffer as much. Conversely, the more males do around the home and with the children, the more pride they experience and the more they credit themselves and others credit them with doing the right thing. One of our female respondents claimed, "Men feel great if they do 10%; women feel guilty if not doing 100%."

Women tend to feel guilty about their professional growth if they are attending to home and family responsibilities; and they tend to feel guilty about their caring/emotional development if they are attending to their professional responsibilities. We have several respondents who claim that at least at certain times in their lives, this guilt has been debilitating. Women scientists married to other scientists publish on average 40% more than women who are married to nonscientists,[3] perhaps because their spouses appreciate their addiction to science and offer moral support. Others may also serve as constructive critics. Almost 80% of married women in the report cited were married to other scientists.

Our questioning leads us to believe that single females or married females without children are less apt to behave in the computerlike mode described. Many enjoy the luxury of uninterrupted scientific thought from early morning to late at night. Whatever the personal life-style experiences, the hours at the laboratory and office are spent in executing details incumbent on all science professionals.

THE FUTURE?

"Women and minorities are encouraged to apply." "Special consideration will be given to dual-career couples, women, and minorities." "We are an equal opportunity

employer." "Send a duplicate application to the Affirmative
Action Officer." "EEO." Familiar? Of course. No one would
think to write and submit an advertisement for a professional position without appropriate reference to standardized statements of justice. Parity in pay, hiring practices, and
advancement is now expected, and more often than not
achieved with gentle reminding.

Present-day issues are more complex. What will *attract*
women scientists who have a family to a specific job or work
environment and will encourage them to *stay*? The following
includes a partial listing:

- Separate but on-site quality day care

- Child care allowances and/or arrangements for attendance at meetings, evening business dinners, research cruises

- Liberal maternity benefits

- Flexible work schedules

- Time off with "stop-the-clock" policy for tenure decision

- Annual family-day arrangements (ideal over the
  Christmas holiday season) encouraging children and
  families to visit the workplace and to know what a
  parent does in that environment

- Dependent care and independent school fee policy,
  pretax dollars

- Health and dental benefits

- Cafeteria-style benefit packages

- Abolishment of nepotism in stated policy and in practice

- Paternity benefits

The Scientist[1] had several articles focusing on the progressive policies at the University of Oregon. This institution can be heralded for many things, among which having nine presidential young investigator (PYI) awardees, of which six are female. This is the highest proportion of women PYIs in the nation. One observer stated:

> These women, their departments, and their institution testify to the ability of science to accommodate the needs of female researchers. Oregon's administrators, department heads, and faculty demand top-notch research. But the institution appeals to women because it is also supportive of and sympathetic to family needs. The departments try to find university jobs for spouses (male or female) of faculty being offered positions and offer flexible schedules to women who are starting families. Oregon's reward, as indicated by its improved citation record and number of prominent scientists, signals that science and families do mix. It's an environment that young professors find very attractive.[1]

SUGGESTED ACTIONS

An excellent description of dual-career couple opportunities and challenges is given by C. Sue Weiler and Paul H. Yancy.[4] We have adopted many of their recommendations:

- Provide institutional options for both females and males to detour for personal reasons, stop-the-clock, and then return to the fast track.

- Work for federal grant/institutional salary support supplement for child-rearing years so that increased technician/technologist time can be supported by female scientists.

- Women who plan to have children should negotiate with the hiring institution *in advance of accepting an academic appointment*, for a year off—one semester during and one semester after pregnancy.[5]

- Assure that as men and women move toward household equity, policies must be developed to ensure that neither gender is a victim of societal and professional perceptions.

- Provide institutional floating positions; a designated pool of positions exists and each is assigned to the individuals rather than departments and are then returned to a central pool.

- Provide institutional prefill positions; hiring of a candidate can occur prior to the time when a regular tenure track position opens up.

- Implement a policy incorporating telecommunication use in an office in the home.

- Establish an institutional Dependent Care Reimbursement Program so that expenses for day care/education of children and/or elders can be paid with pretax dollars.

- Avoid scheduling any action that bypasses mothers or punishes scientists who are parents, particularly evening seminars or dinner meetings, cocktail decision sessions, or weekend meetings (now popular due to the reduced stay-over-Saturday-night fares).

SUMMARY

Present-day issues in the world of science and the business of doing science are complex. Each day scientists are confronted with multiple priorities. These priorities can be confounded by the parenting role that still falls disproportionally on women. Daily considerations include prioritizing the science professional, the parent, the partner, the humanist, the naturalist, and the self.

It is clear that the capability to manage multiple priorities with confidence is one key to advancement to greater responsibility and through the hierarchy of academia or governmental agencies—whether for females or males. The simple recommendation is therefore to segregate the professional and parental aspects of one's life. With an extremely reliable professional support staff and a strong and generous parental support group, this is possible. Many women have successfully done this, individually mastering their situations. But the goal for a rewarding career should be an integrated life-style.

Advances must be achieved at the population level. Every women who selects a career in mathematics, science, technology, or engineering should be able to depend on a built-in system that supports a solid network of professional support staff plus child care, school, and elder care options that provide a nurturing environment for their offspring and their elders. We must work diligently to provide these opportunities and to leave this legacy. First-quality institutional, state, and national day-care programs for all must become an expected reality.

Indeed, present-day issues are complex. With appropriate support, parents can enjoy both family and professional life. The demands are high; the rewards are high. The race is one of performance and endurance and burnout is common. The most important defense is to *reserve some time daily for yourself.*

REFERENCES

1. Elizabeth Pennisi, Flexibility, balance draw women to the University of Oregon, *The Scientist* 4(20), 7 (1990).
2. A. H. Koblitz, Career and home life in the 1880s: The choices of mathematician Sofia Kovalevskaia, in *Uneasy Careers and Intimate Lives: Women in Science 1789–1979*, eds. Pnina G. Abir-Am and Dorinda Outram (Rutgers University Press, New Brunswick, NJ, 1987) , p. 120.
3. C. Sue Weiler and Paul H. Yancy, Dual-career couples and science: Opportunities, challenges and strategies, *Oceanography* (November 1989), pp. 28–31, 64.
4. Jonathan R. Cole and Harriet Zuckerman, Marriage, motherhood and research performance in science, *Scientific American* 255(2), 119 (1987).
5. Barbara Spector, Women astronomers say discrimination in field persists, *The Scientist* 5(13), 21 (1991).

CHAPTER 5

EMPLOYMENT, UNDEREMPLOYMENT, AND UNEMPLOYMENT

The continuing scarcity of women in professional scientific positions—causes and remedies; part-time scientists; the many categories of underemployment; unemployment.

The range of issues to be addressed in a chapter concerned with "employment" is extensive enough to warrant an entire book just on this topic. For the present, though, we have selected some areas that are of particular significance to women scientists. They are:

- Relative numbers of women scientists in academic, governmental, and industrial research organizations
- Part-time scientists
- Kinds of underemployment
- Unemployment

Competition for professional jobs is and will continue to be severe, especially in an era of declining student enrollments, increasing costs, and continuing weakness in government funding of research. Employment thus becomes a highly selective process, with the best positions going (ideally) to the best qualified. The realities—especially for women scientists—may be less than ideal, and will be explored in this chapter.

RELATIVE NUMBERS OF WOMEN IN SCIENCE WORKPLACES

It is a common perception, supported by data from many academic, governmental, and industrial research organizations, that the relative numbers of women in science have increased remarkably in the past three decades. Examined more closely, though, it can be observed that as recently as 1987 women accounted for only 13 *percent* of all scientists, and even these were heavily concentrated in the social and behavioral sciences.[1] Furthermore, it appears that the major increases in numbers occurred during the late 1960s and the 1970s, and that the rate of increase actually diminished during the 1980s. Statistics from a number of institutions indicate a recent loss in momentum. At one large Eastern state university (Maryland), for example, 8 percent of the full professors in 1982 were women. Six years later, in 1988, that figure was virtually unchanged, and the total female faculty had increased by only 0.5 percent in the intervening time. Similar evidence of stasis—of leveling off in the 1980s—can be seen in the numbers of women entering science when compared with the previous two decades.

This trend is also evident in specific fields. An example is presented from the field of oceanography[2] (Figure 11).

The reasons for this rate change are undoubtedly complex, but certainly some of the principal components are decreased vigor on the part of political administrations in all equal employment efforts, combined with reduced federal and state funding levels for many agencies and industries. Funding for nonmilitary scientific research has been particularly vulnerable.

One consequence of reduced governmental emphasis on equal employment seems to have been abatement of "forced" hiring of women scientists because of institutional programs to raise the proportion of female professionals on the staff. Those programs were usually aided by pressure from an aggressive dean or vice president (often but not always a woman). They attempted to respond to the obvious minority status of women in science, but progress in improving the female:male ratios among university faculty members in general has been and remains remarkably slow.

As an example—not at all atypical—of the continuing disparity in relative proportions of men and women at each academic rank, a report published recently by one of the largest of the Eastern state university systems (Florida) disclosed the comparative numbers for 1987, shown in Figure 12. Parity seems to have been achieved only at the introductory level, with increasing disparity at the (tenured) higher ranks. Other possible reasons for the diminished rate of increase of women faculty include severe financial problems in a large number of academic institutions, with consequent depressing effects on hiring and tenure awards. Whatever the causes may be, the long-term effects are to slow the rate of acquisition of women faculty as role models

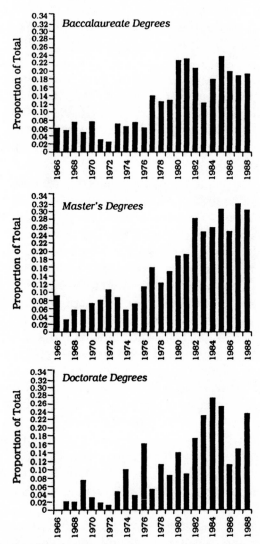

FIGURE 11. Proportion of various degrees awarded to women from 1966 to 1988 in the interdisciplinary field of oceanography. SOURCE: Luther Williams, 1990.[2]

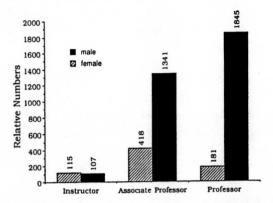

FIGURE 12. Relative numbers of male and female faculty members of various academic ranks in Florida's state university system in 1987. SOURCE: Cathy Shaw, *Miami Herald*, January, 1989, p. C-1.

for female undergraduate and graduate students and to slow the overall rate of movement of women into academic power positions.

An interesting and no doubt significant recent phenomenon accompanying declining female enrollments in science is a counterbalancing increase in foreign (particularly Asiatic) students, teaching assistants, and faculty who—as was pointed out by one senior university administrator—bring gender biases from *their* cultures.

The observed changes are reflected in recent faculty hiring practices. American colleges and universities, still claiming to be "equal opportunity employers," have adopted policies that make race or ethnicity overriding factors in hiring new personnel, with a concurrent deemphasis on women, except as they represent minority races or ethnic groups. Bonuses are offered by some institutions to departments that hire minority faculty members; some state systems (California, for example) now require that a specific percentage of all new faculty of community colleges be

minority; some schools hire minority doctoral candidates
and subsidize costs of completion of degree requirements.
Women seem to be overshadowed in this new racial/ethnic
thrust.

Another recent trend is the choice of careers in computer science by women with science or engineering training. Reasons given are that a rigid mathematics background
is not necessary to enter the field, thereby making it more
accessible to those with social sciences training (psychology,
anthropology, economics); and that it is a relatively new field
with a hoped for lower prevalence of biases against women
than might be found in other science and engineering fields.
The realities, as discussed by Peggy Schmidt[3] in a *New York
Times* article, are that women are paid well, but less than
men for the same jobs, and that they are often relegated to
jobs such as client training and testing and to detail jobs in
software production. Despite these limitations, Schmidt
reported that women comprise one-fourth of all software
specialists—a higher proportion than in other fields except
for biological sciences, with one-third the work force female
(but not all working as professionals). Disparities were also
seen at managerial levels in computer science: Schmidt
reported that, on average, in 1986 the highest-paid woman
manager earned less than the lowest-level male manager.

Another indicator of greater participation of women in
mainstream science and engineering has been a doubling
in the past decade (1977–1988) in the percent of U.S. patents
issued to women inventors. Patents have of course long been
a part of industrial science, but have only recently been
pursued aggressively by the universities, where the tradition had formerly been to publish but not to patent new
research findings. In a recent article, Kathryn Phillips[4]
reported that in 1988 women received 5.6 percent of U.S.

patents—still a small figure, but better than the 2.2 percent reported for 1977. She attributed the increase to the greater representation of women professionals in science and to policies by large science-based industries such as AT&T and Dow Chemical, encouraging the hiring and promotion of women scientists in significant numbers. One interesting conclusion reached was that "Once men and women are in the same scientific environment, there seems to be no difference in what they invent . . ." (As an interesting postscript to this discussion of women inventors, the U.S. Department of Commerce announced in April 1991[5] that the first women inventor, Gertrude Elion, had been elected to the National Inventors Hall of Fame, taking her place with Edison, Bell, and 92 others in this elite assemblage of talent.)

A recent analysis by Rosabeth M. Kanter[6] suggested that, as a general rule of thumb, a minimum critical mass of females was needed to even begin to achieve any semblance of a balance of power in a previously male-dominated organization. Implicit here, in addition to mere numerical presence, would be gradual assumption by women of key roles in evaluative and decision-making bodies as seniority is achieved. Numerical advances are not enough, however. To compete successfully in the present system of science, women have to be at least as aggressive and innovative as their male counterparts; many women do not seem to accept this as reality.

Among the available innovative approaches mentioned in an excellent review article on the status and needs of women scholars by Mary L. Spencer and Eva F. Bradford[7] were:

- starting a new research and development or consulting organization, outside the academic environment

- organizing an information network for women scientists

- beginning new women-administered research projects

- organizing technical workshops and conferences

- establishing support groups and conducting workshops (for men as well as women) on changing attitudes

- accepting short-term appointments with agencies and organizations that administer research funds

- actively soliciting research collaboration with male colleagues

- increasing involvement in the affairs of professional organizations

These and many other approaches listed by the authors were categorized as changes in personal and interpersonal strategies, as well as institutional/societal changes. Spencer and Bradford also point out, quite correctly, that "The burden of change must be shared by everyone within the academic and scientific community."[7]

PART-TIME SCIENTISTS

In confronting the issue of part-time scientists, the discussion often reverts to science versus the home, which is a central one in the thinking of some women professionals. It must be pointed out that there are other reasons for part-time efforts, and that men and women alike seek part-time employment. Some want to pursue further studies in science or mathematics. Some want to pursue studies in other areas including the arts. Others run part-time consulting businesses.

We should ask immediately, "Is the designation of 'part-time scientist' a contradiction in terms—is it possible to be anything less than fully committed to the practice of science?" Predictably, answers tend to vary with the gender of the respondent. Many male scientists look on science as a full-time all-encompassing occupation. For some of them, absorption is complete, leaving little if any time for other pursuits. One humorous description entitled "The Addictive Personality" appeared recently in the journal *Science*[8] and is reprinted here:

> SCIENCE: Dr. Noitall, you are the world's greatest authority on addiction, the seer that everyone consults, the man who got Sherlock Holmes to kick his cocaine habit.
>
> DR. NOITALL: A vast understatement of my true worth.
>
> SCIENCE: Could you describe the addictive personality?
>
> DR. NOITALL: An addictive person is one who has a compulsion to behave in ways that his or her family members consider detrimental to their interest. An addictive person will frequently conceal the extent of his addiction, will lie to his family about it, is immune to logical arguments to correct the errors of his ways, and forgoes income that would require abandoning the addiction.
>
> SCIENCE: Are we talking about a dope addict or alcoholic?
>
> DR. NOITALL: No, I am describing a scientist. It is well known that work habits of scientists are addictive, leaving their spouses in tears, their children pleading, "Come home, Mommy (or Daddy)," and involve long hours in hostile instrument laboratories or cold rooms, exposed to noxious gases and radioactivity—conditions that no sane person would choose.
>
> SCIENCE: But surely these individuals are paid handsomely for undergoing these hazardous conditions.
>
> DR. NOITALL: This is the peculiar paradox. The profession is

poorly paid because there are hundreds of applicants for every good position. Because of the psychic income that is exploited by our oppressive society, a scientist will accept pay that would make a movie star weep.

SCIENCE: But many of these individuals are academics who have the advantages of long summers off and light teaching loads.

DR. NOITALL: Academic freedom is the freedom not to take a vacation. Far from taking summers off, these individuals would rather develop films in the darkroom than sit on the beaches of Waikiki.

SCIENCE: But surely these individuals have a record of stable homes, paying their bills, and other behavior not typical of an addict.

DR. NOITALL: That depends on how you define good behavior. These individuals tend to curl up with a copy of the *Physical Review Letters, Journal of the American Chemical Society*, or *Journal of Biological Chemistry*, rather than doing household chores or acting like good Americans who stay glued to the television set.

SCIENCE: So far, however, you have merely described an individual who works to keep his job.

DR. NOITALL: No, these individuals are definitely masochistic. They volunteer to serve on review panels that send them hundreds of incredibly detailed project proposals which must be read and evaluated. They sit through endless thesis defenses, volunteer to edit journals, and serve on visiting committees for other schools when they have too much to do at home. They then complain bitterly that they are too busy.

SCIENCE: It is apparent that these individuals could do well in other occupations?

DR. NOITALL: They are addicted to scientific logic, which makes it impossible for them to act like a trial lawyer who sues the city of New York for negligence when a drunken man falls off a subway platform, or a politician who claims one can increase services and pay lower taxes, or a movie star who testifies before Congress on carcinogens but does not know the difference between valium and valine.

SCIENCE: Is there any behavioral characteristic that can explain this obsessive conduct?

DR. NOITALL: Basically scientists have failed to grow up. They are all children, eternally curious, eternally trying to find out how the pieces of the puzzle fit together, eternally asking Why, and then irritatingly asking Why again when they get the answer to the first question.

SCIENCE: But can't this addiction be cured by some new program of drugs and therapy?

DR. NOITALL: There is no evidence of hereditary characteristics of family environment to produce a scientist; therefore we have few handles on the potential cures, but the most glaring fact is that society cannot afford to cure these individuals. Their obsession is responsible for most of the progress of mankind and therefore the last thing we need at this moment is to turn these addictive scientists into well-adjusted television watchers. It is well worth giving them the tiny bit more money they need to stay addicted to science and to attract new compulsive personalities to their work. Society is as addicted to scientists as scientists are addicted to science.

Much has been written about obsessive or workaholic scientists—usually but not always males. Searching questions have been asked (and have usually received inadequate answers):

- Is total commitment to science an adequate substitute for many other meaningful things in life—home, family, friends, hobbies?

- Is this behavior a sign of true dedication to a calling, or a compulsion out of control?

- Is science, with its abnormal and often extended work weeks, a retreat to avoid elements of a normal life?

- Is this "replacement" or "substitution" policy—in which science becomes overriding—more common among male scientists than among females?

As might be expected, responses from male and female scientists varied to some extent—within and between genders. Some men expressed genuine pride that their occupation was so all-absorbing, regardless of the consequences; others—more than the first group—admitted, sometimes a little ruefully, that the practice of science probably took more of their time and energy than it should, leaving little time for family or hobbies. Still others—probably a majority—denied such total involvement, stating that science to them was a daily job, albeit a good one, and that it had time priority, but not to the exclusion of other interests.

The pattern of responses from unmarried women scientists was generally similar to that of the men. The notable exceptions were from those women scientists who were married, based principally on the joy and demands of home and family. Women with such external responsibilities denied that they were any less dedicated or less committed to science than men *for that part of their time that could be allocated to professional activities.*

Family involvements sometimes forced the decision to accept part-time positions, or modified work schedules, or periods of leave, but this was not deemed by women respondents to demonstrate lack of interest or only casual commitment to science. Those involvements did, however, produce "deflections" in career progression, often with negative impacts on rates of publication, promotions, and tenure awards. Frequent references were made in questionnaire responses to the persistence of unfair perceptions among colleagues and administrators that less than full-time schedules implied less than complete commitment to the profession—a perception strongly denied by many women scientists. They point out that it isn't only the *amount* of research that is done, but also the *quality* of the work that counts. A few robust and highly significant publications on challenging research topics placed in the best journals are far more valuable than several smaller, less scholarly contributions. Moreover, some scientists require 100-hour work weeks to be even marginally productive, whereas others can operate at 20 percent of capacity and still accomplish significant work. It is possible for both females and males to think conceptually during the commute to the child care center, to jot down important points for the manuscript draft while the baby naps, and to design experiments while mowing the lawn.

Why should anyone have to be only half a human being to do good science? With hard work and organization, a career in science can still leave room for a fulfilling personal and family life. It may be, as Vivian Gornick[9] believes, that

> the idea of worshipping at the shrine of the "Scientific Enterprise" is being replaced by a more rational view that the enterprise is after all a human invention, born of a basic need of the human mind to make sense of the universe; that it is not greater than we are; and that none of us should be sacrificed on its altar, whether we are male or female.

Underlying any discussion of career versus family should be some reference to what seems to be a major shift in women's attitudes about scientific vocations. Some observers have noted a sharp change beginning in the 1960s in women's *expectations* of careers in science. New graduates before that would expect to choose between career and family, possibly with a vague and usually unrealized intention to return to science later, if the family option was selected. Since that time, more and more women science graduates have been insisting on *both* options—career and family—*simultaneously*, with institutional and family arrangements (day care, flexible schedules, spousal support) that make combined careers feasible, with full-time commit-

ment in science a reality. It is also interesting that more and
more male science graduates have been insisting on both
options—careers and family.[10] The rewards of these enlightened parental partnerships are being realized by mother,
father, and children.

Despite these great expectations, many of today's
women scientists still feel that they have been forced to
choose between starting a family and pursuing a professional career in a specialty that may have only a limited
number of very competitive openings. They have concluded
that women in science cannot succeed if they insist on being
"women" in the societal sense of the term; this is evident
from the observation that many women scientists are unmarried, or divorced with no intention of remarrying, or
married with no children.

This state of affairs was examined recently (1987–1988)
in a survey by a committee of the American Phytopathological Society. Some significant conclusions were:

- Continuing difficulties in integrating professional and
  personal lives, from graduate school throughout their
  scientific careers (Figure 13).[10]

- Continuing inflexibility of most academic careers
  combined with the reality that the most demanding
  (pretenure) work years coincide with peak reproductive years.

- Continuing perception by colleagues and administrators that commitment to science is no longer taken
  seriously if women marry and/or have children.

- Continuing lack of or inadequate child care opportunities and absence of or weak maternity leave policies.

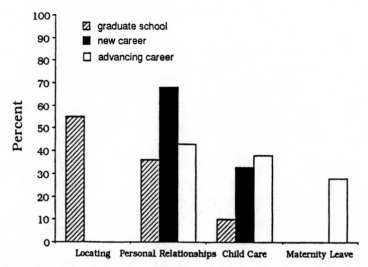

FIGURE 13. Difficulties encountered by women plant pathologists in integrating professional and personal lives [from Marlys Cappaert *et al.*, *Phytopathology News* 23(6): 75–76, 1989].[10]

Some suggested remedial steps that could be taken by the scientific community included provision of child care at professional meetings, promotion of national family leave legislation, and encouragement of the development of part-time positions.

To balance this discussion of the study by the American Phytopathological Society, it should be pointed out that two-thirds of the women queried were satisfied with their rank and salary and that the other one-third had complaints that focused on factors in addition to those involved in the home versus career issue. Some of these were:

- Continuing hiring discrimination in which women were either discouraged from applying for positions or were not selected because the position was "not for women." These practices are of course illegal, but apparently persist.

- Continuing discrimination in the form of salary differentials, delays and obstacles in career achievement, and exclusion from professional activities.

- Continuing lack of role models, feelings of isolation, and lack of recognition or respect for qualifications and job performance.

When combined with problems associated with integrating professional and family lives, the totality of career barriers can be perceived to be, and can actually be, overwhelming and discouraging for the woman scientist.

KINDS OF UNDEREMPLOYMENT

In the previous section we considered some aspects of science done on a part-time basis—usually by women professionals and usually for compelling reasons. Other forms of underemployment—those that fail to engage the full range of professional competence—exist, and women seem to be victims in many instances when other more desirable options are blocked.

1. *The part-time instructor hired to teach specific courses.* Salaries for these positions are usually abominable, proportionally far below those of full-time faculty members, if examined on a contact hour basis. Such jobs are nontenure track, with short-term contracts

and few if any fringe benefits. They are often clustered in junior colleges and some urban universities.

2. *The research associate.* This kind of position is undoubtedly the most common cause of underemployment among women scientists, so much so that it will be given detailed treatment in Chapter 6 on "Scientific Support Roles." All we will say here is that, with rare exceptions, research associate positions should be accepted only *in extremis*, and never for any significant length of time. (One possible exception to this might be a short postdoctoral appointment in the laboratory of an outstanding leader in a specialty area.)

3. *The "courtesy" appointment.* This pernicious kind of appointment is closely related to (and may actually be in the form of) a research associate appointment, but usually has different motivation. It persists in instances in which successful recruitment of a scientist with particular skills or significant reputation can only be effected if the spouse (also a professional) is offered a paid part-time or full-time nontenure track research or teaching position in the same or a different department. Too often the courtesy appointment is offered to the female member of the pair.

4. *The junior college trap.* Junior or community colleges can serve an important post-high school role for specific technical training, but they are at the bottom of the higher education barrel in terms of professional advancement and job satisfaction for competent scientists on their faculties. These institutions

constitute a clear menace to upwardly mobile professionals; research is not encouraged and is sometimes not even condoned; student contact hours are crushing; and teaching experience at this level is often not equated with that at four-year college or university levels.

5. *The small privately funded liberal arts college.* Literally thousands of small colleges exist in the United States. They are primarily teaching institutions, some marginally financed with commensurate minimal faculty salaries. Because of their sheer numbers, they are potential sources of entry-level jobs for many new professionals, but they can be dead ends for the unwary, since, as with the junior colleges, teaching loads are usually excessive and until recently faculty research has had very low if any priority with their administrators.

6. *Industrial laboratories.* Industrial research positions can offer satisfying careers for a highly selected group of professionals. Pitfalls exist, though, especially for the entry-level scientist. Often the position title and description can be interpreted as "professional" but the actual duties and responsibilities are those of a technician or technologist. Also, the narrow mission orientation of the research may leave little room for individual initiative, and publication of findings is often strictly controlled or discouraged.

7. *Government science bureaucracies.* Eager well-qualified scientists may be tempted to accept positions with state or federal regulatory agencies that have subsidiary research functions (organizations such as state

departments of natural resources or the United States Environmental Protection Agency). The job descriptions may suggest research activities, but incumbents are usually quickly absorbed into bureaucratic functions and become totally immersed in a river of paperwork supporting regulatory actions, never to emerge and never or rarely ever to practice hands-on science.

It is obvious that most of these forms of underemployment of highly trained scientists have few if any overt gender distinctions (except for the ubiquitous research associateship). It is also obvious—at least to us—that because more women, proportionally, are at entry or junior professional levels, they are more likely to populate the marginal kinds of positions just described. However, the powers that be are not pushing to disturb the status quo.

UNEMPLOYMENT

Women scientists are subject to periods of professional unemployment, for reasons less frequently encountered by male counterparts. Some of them are:

- Being "forced" to take an interim nonprofessional job because of a spouse's move to a new position and lack of professional niches in the new location.

- Too-long residence in a research associate position funded by someone else's grant, and the termination of that grant.

- Too-long residence in a nontenure track position

whose continuity depends on external institutional funding sources, which may dry up or disappear.

- Confronting a university or agency administration because of perceived unfair unemployment practices, with usual long delays—sometimes with associated unemployment—before resolution of legal proceedings.

- Nonrenewal of short-term faculty or agency contracts, because of cutbacks dictated by institutional financial problems and not because of job performance.

Unemployment, for whatever reason, is of course a very unwelcome condition. Apart from the absence of a salary, it can erode feelings of self-worth and it usually represents a valley or a plateau in career progression. Unemployment is destructive to the ego if it results from inadequate job performance, but it is especially frustrating if it results from perceived or actual factors other than lack of professional competence or productivity.

In rare instances, unemployment for women scientists may also be the result of conscious choice, usually the consequence of an amalgam of several underlying factors, and often with a single precipitating event, either personal or professional. Some of the factors can be:

- Perceived discrimination in promotions, salaries, or tenure decisions

- Rebuffs, harassment, lack of support, or open hostility from colleagues or administrators—being "iced out"

- Minor slights or put-downs, couched as humor but not humorous

- The impostor syndrome—a feeling of not being adequate for the job

- Disenchantment with a chosen specialty area

- Family pressures, especially lack of support from a spouse/partner

- Divorce or death of spouse/partner

For some, a point is reached when the conclusion has to be "I don't need all of this grief," and the decision is made to leave science altogether and never to return.

An example of a decision to leave science completely because of an accumulation of negative events was told to us, rather forcefully, by friends of a former staff member of a large private southeastern university.

Dr. Angela Marks had been employed for more than a decade as a research associate and had made substantial contributions. A faculty appointment had not been forthcoming during all that time, principally because several earlier appointees (all males) occupied the available positions in her specialty area. These faculty members probably felt threatened by her capabilities and were openly hostile, so that the departmental atmosphere was unpleasant most of the time. The woman scientist did not want to move elsewhere because of family commitments and community involvements.

The precipitating event occurred at a national society meeting, where a paper that she presented was treated unnecessarily harshly by her departmental "colleagues" who were in attendance. One of them (who probably saw his "turf" being threatened) went so far as to circulate a letter suggesting that she be denied future membership in the society—a stunningly unethical act. This was the "Who needs this?" decision point for her. She left science soon after that episode and is now a successful yacht broker.

Another example of a decision to leave science altogether was described to us by Dr. Carla Foster.

> Dr. Foster, a widely published career researcher in astronomy, was in her late thirties, married, and blissfully became pregnant and delivered a healthy baby. She returned to active full-time research within 3 months of the baby's birth. Her assistant had kept her laboratory effort active and productive during Dr. Foster's absence. One year after her return, she was torn and eventually decided to throw in the towel, to quit science ("writing proposals!"), and to pursue her hobby, pottery, which she could do in her home studio. Male superiors, colleagues, and subordinates all felt that the decision was irrational and menopause related. In an official letter to a funding agency, her supervisor (male) attributed her decision to a "health problem." Dr. Foster had no female superiors, but her female colleagues and subordinates all applauded her and envied her ability to make an unexpected decision. Dr. Foster became a folk hero.

SUGGESTED ACTIONS

Some approaches to increasing the female/male ratio in university faculty populations and to ensuring equal employment practices include:

- Developing a vital and visible women's faculty association to promote hiring and promotion of female professionals
- Preparing for public release of annual statistics on the proportion of women faculty members, with analyses

of short- and long-term trends, based on rank and
length of service

- Preparing stated and highly specific university and
industry policies on hiring and promotion of women
professionals, followed by acceptance and implementation of those policies by the administration, all under the scrutiny of a watchdog subcommittee of the
women's faculty association

- Identifying women administrators in the university
and industrial hierarchy (department heads, deans,
vice presidents, chairpersons) who are willing to accept the additional responsibility of insisting that a
woman's perspective be applied to decisions that concern hiring and promotions

- Communicating with and supporting *male* administrators in the university and industrial hierarchy who
have demonstrated genuine and not merely *pro forma*
interest in ensuring equal hiring and promotion practices

- Maintaining a vocal female presence in deliberations
of faculty senates or counterpart internal bodies with
advisory or decision-making authority.

SUMMARY

Women are dramatically underrepresented in the professional scientific work force, making up less than 15 percent of all scientists. Some increase in numbers took place in
the 1970s, but the rate of improvement was not sustained

during the 1980s. Reasons offered for the continuing disparity in numbers of female versus male professionals include: decreased vigor in implementing equal employment policies, sharply reduced funding for scientific research, financial difficulties at many academic institutions, lack of programs to attract girls to science courses at elementary and secondary school levels, and avoidance of hard science majors by female college students. Women are particularly underrepresented at the senior, tenured faculty levels, and parity with men exists only at the most vulnerable introductory levels.

The disproportionate male:female ratio in science is exacerbated by the reality of underemployment for many female professionals: in part-time college teaching positions, as research associates, as junior college instructors, as technicians in industrial research laboratories, or as minor bureaucrats in government agencies. Since proportionally more women than men are at entry or junior professional levels, they (the women) are more likely to accept such marginal jobs.

It could be argued that any kind of professional position is preferable to unemployment, and women scientists may be unemployed for several gender-related reasons: their husbands may accept positions in remote locations without professional niches for them, their research associateship may be terminated when support funds dry up, or they may become totally disillusioned by persistent gender discrimination in their employing organization. Recent analyses have confirmed repeatedly the slow progress being made in augmenting the proportional numbers of women in science and in distributing them more equitably throughout academic hierarchies, and remedies have been offered, especially programs to keep women in educational pipelines, to improve the academic climate for women, and to achieve salary and prestige levels found in other professions.

REFERENCES

1. Jonathan R. Cole, Preface to the Morningside Edition, in *Fair Science: Women in the Scientific Community* (Columbia University Press, New York, 1987), pp. xiii–xx.
2. Luther Williams, Human resource trends in oceanography, *Oceanus*, 33, 12 (1990).
3. Peggy Schmidt, For the women, still a long way to go, *New York Times, Special Section: The Market* (April 1986), pp. 14–15.
4. Kathryn Phillips, U.S. Patent and Trademark Office study finds more inventions credited to women, *The Scientist* 4(20), 24 (1990).
5. Oscar Masten, First woman inducted, National Inventors Hall of Fame, *Commerce People* (April 1991), p. 3.
6. Rosabeth M. Kanter, *Men and Women of the Corporation* (Basic Books, New York, 1977), pp. 208–209.
7. Mary L. Spencer and Eva F. Bradford, Status and needs of women scholars, in *Handbook for Woman Scholars: Strategies for Success*, eds. Mary L. Spencer, Monika Kehoe, and Karen Speece (Americans Behavioral Research Corporation, San Francisco, 1982), pp. 3–30.
8. Daniel E. Koshland, Jr., The addictive personality, *Science* 250(4985), 1193 (1990).
9. Vivian Gornick, *Women in Science: Portraits from a World in Transition* (Simon and Schuster, New York, 1983), pp. 158–161.
10. Marlys Cappaert, Mary Powelson, Virginia Stockwell, and Kathy Merrifield, Women in plant pathology survey: Part II, *Phytopathology News* 23(6), 75 (1989).

CHAPTER 6

SCIENTIFIC SUPPORT ROLES
Associates, Technicians, and Assistants

Being in science but not of science; the indispensable research associate; the necessary laboratory technician/ technologist; the research assistant; why so many women?

The discussion in this book thus far has focused on women as professionals, as degree-bearing practitioners of the complex art of doing good science. We decided early on and quite arbitrarily that necessary qualifications of a scientist were (1) *the possession of an advanced degree* in one of the many technical specialties—in the "hard" or "soft" sciences—and (2) *employment* in a position that makes use of that background. This seemed at the time to be a logical decision, and it has carried us this far. But periodically in the discourse we have encountered another large category of women in science—those in *supporting* roles, making specialized contributions to the success of projects (and project leaders), but usually doing so at a "subprofessional" level, often

111

without benefit of advanced degrees. So it seems necessary, midway in this examination of women in science, to give some visibility to all of those in roles that do much to make science function: the research associates, the laboratory technicians, and the laboratory assistants in numerous categories.

There is a problem, though, with this or any attempt to insert people into neat pigeonholes, even if only for discussion purposes. They just don't all fit. The range of supporting activities in science is too broad—all the way from a highly trained degree-bearing research laboratory supervisor to a competent public relations/proposal-drafting specialist, to a skilled laboratory technician or technologist, to a part-time aide who spends evening hours entering experimental data into sophisticated instruments. But a start has to be made, and we propose to isolate and discuss a few major categories: (1) the research associate with one or more advanced degrees, (2) the laboratory technician/technologist, and (3) the research assistant or aide.

THE RESEARCH ASSOCIATE

Of all the women in supporting roles, the research associate is the most difficult to characterize adequately. Often employed in jobs with marginal professional status and no tenure, despite possession of one or more advanced degrees, such women present varied histories. Some are perennial support staff members (and often former graduate students) for authority figures; they may feel fulfilled by the continuing contact with a mentor, or they may be just reluctant to assume the role of independent investigator, or they may not have been able to find an acceptable position elsewhere. Others may be postdoctoral appointees who deliberately seek out the best research people in their area of interest and join their research group for one or more years. Research associates often form the real substance of team competence and may be responsible for much of the productivity of the research group. Eventually, some achieve independence, but sometimes at a cost in delayed career accomplishments.

A much larger segment of the research associate category is drawn from the pool of women who move because their spouses do and who may not always find appropriate faculty positions in the new location. They then take temporary, borderline professional or clearly subprofessional jobs, working in the laboratory on someone else's grant. Time goes by. Some escape by finding faculty slots; some enjoy the relative lack of stress and flexible hours, and continue in a succession of precarious year-to-year temporary positions and some eventually disappear from science altogether, becoming deflected or derailed.

Examples of the research associate syndrome can be found in almost every academic institution. We have selected two.

> Dr. Annette Munk and her husband, John, were members of the same department of a major Eastern university. He had tenure, she was first a postdoc, then a research associate, although both partners were approximately the same age. Dr. Annette Munk was the perpetual research associate in the eyes of everyone in the department. She and her husband were viewed as equals in the grant and contract arena based on their success rate with external funding. Colleagues praised them for their joint publications, readers giving both authors equal credit. The Drs. Munk found the inequity at their home university intolerable and applied to other universities. They received several offers, but always with a similar male:female imbalance. It

was only when both were offered tenure track positions that they uprooted and moved.

Dr. Melissa Morris and her husband both returned to a prestigious Midwestern university from federal agency jobs in Washington, D.C. While the husband was offered a tenured position, his pregnant wife was offered a courtesy research associate position. In subsequent years, Dr. Melissa Morris regularly applied for tenure track positions as they became available in the department. Rarely was she even interviewed or shortlisted. Never was she hired into one of these positions. Eventually, a prestigious government agency offered a high-powered job back in Washington, D.C. to her. Only then with the threat of "loss" did the department upgrade her status and position.

In discussions with women in these long-term research associate jobs, an undercurrent of unease can be detected. Some feel that they are living on the margins of science, marking time until a better opportunity appears and recognizing that the delay in career development is damaging. Others find that the scientific contributions made in their present positions are ignored altogether or are attributed to others—the "perennial junior author syndrome." Still others feel moderately satisfied just to find employment at any level in the field for which they were trained, even though they had expected more (and still expect more).

Research associates usually bring specialized skills to a project, and frequently provide critical continuity when the project leader—often a male—is off on the conference/ symposium circuit or is too busy writing technical papers or new grant proposals. Research associates often supervise a cadre of technicians and assistants, and customarily are also responsible for the care and feeding of graduate students, as proxies for the often-absent faculty member directing the project. Research associates are untenured, underpaid, often unrecognized, and usually unhappy, although exceptions may be found. They do fill a vital role in the enterprise of modern scientific research, and for some the job, if not prolonged, serves as a valuable internship under a leader in a specialty area.

An extreme example of the almost inexplicable persistence of the "research associate syndrome" in a changing society was described to us in detail by a colleague of Dr. Harriet Traverse, a good scientist with a long history of successful teaching at a major university. However, she was also the wife of the eminent scientist, Dr. George Traverse. With her husband's last move ten years ago to a privately funded research institute in the Southeast, she was offered a desk and a nonsalaried "cooperator" position so she could work on a peer basis with her spouse (her area of expertise is similar to his), an offer that she accepted on a temporary basis. Recently, Dr. George Traverse died. The institute director, instead of immediately appointing Dr. Harriet Traverse to the position, merely reassured her that she could continue in her unpaid cooperator job and then proceeded to advertise and fill the position with a recent Ph.D. at an entry-level salary. The important fact to note here is that such arrangements and such inequalities do exist *today*, despite all the clamor about equality.

A strikingly similar example of the invisible widow as long-term research associate—with a happier outcome—was recounted to us during an interview.

Dr. Ann Heinik worked closely with her husband who was director of a small laboratory. When her husband suddenly died of a heart attack, Dr. Heinik was encouraged to complete the research and teaching commitments of the two of them. This she did. At no time was there any recognition or acknowledgment of her work within her own institution, although colleagues worldwide held her work in high regard. The status quo continued for over 10 years and was tolerated by Dr. Heinik.

No equity was suggested until a female director was assigned to the laboratory. The new director found the perpetual
research associate position intolerable and suggested that a
tenure offer from another institution would be necessary to
evoke a counteroffer for promotion and tenure at her home
institution. Within two weeks an external tenure offer was
made. Within four weeks an internal tenure and promotion offer
was made with notable split decision. The offer was accepted.
No one discounted the value of Dr. Heinik again.

These examples of the persistence and evils of long-
term research associateships for women scientists force us to
ask some hard questions about the system that fosters (or at
least condones) them. Are women scientists less valued than
men on their own turf? Are women scientists automatically
undervalued if they are brought in as part of an arrangement
to recruit their husbands? Is this just another illustration of
the idea that "no one wants you unless someone else does?"
The temptation is to answer yes, yes, yes, to all these questions, but such a response would be too simplistic. Each
case has its own nuances; institutions and administrators
vary in their attitudes toward such positions; personal motivations are complex; and changes in the way science is
conducted are always made very slowly.

THE LABORATORY TECHNICIAN/TECHNOLOGIST

Laboratory technicians/technologists have been and
continue to be predominantly women. Like the research
associate, they provide critical skills, but at a more technical
level. Their previous training can be quite variable, ranging
from undergraduate degrees in science, to two-year associate degrees from junior colleges, to short-term specialized

courses in laboratory techniques, to on-the-job experience with someone already possessing the required skills.

Laboratory technicians may be very competent in what they do, but their progress in science can be forever blocked by the absence of requisite degrees. Some upwardly mobile technicians may take advantage of an academic environment by acquiring those degrees through part-time studies, but many accept the role for what it is and enjoy being very good at what they do. Not unusually, women technicians have major family responsibilities and view their jobs as a way to earn a steady income in a work environment that is usually more pleasant and more stimulating than most.

An artificial but workable distinction is often made between the professional and the technician beyond the possession or lack of advanced degrees. Professionals plan and organize research, and in addition analyze and publish the results of that research. Technicians carry out the more mechanical aspects, under the nominal supervision of the professional, but are not required to *think about* the more conceptual basis of the work they are doing. This is not the true state of affairs in many laboratories, however, since good technicians can make valuable contributions to the design and implementation of studies in which they are involved on a daily basis. The term technologist is favored over the term technician, to denote that the individual is engaged in the study of the techniques utilized. This is in contrast to the professional who is engaged in the study of the scientific question at hand.

Intelligent principal investigators invite participation of technologists in all aspects of the research, developing a genuine team spirit. Everyone in the group is aware of the objectives of the study, and all feel welcome to provide suggestions as the work progresses. Handled properly, this can be considered as effective research management, but it can lead to abuses too. One problem that surfaces repeatedly is that of *appropriate recognition* of the contributions of all participants in the study. Technicians/technologists especially can feel that they have been ignored or underrecognized when the results are published if they are not accorded co-authorship and if, in their estimation, their contributions would seem to warrant such recognition. Knowledgeable project leaders can try to forestall such feelings by clarifying at the outset of the study the expectations from each participant and redefining, if necessary, precisely what the perceived role of a technologist is. This procedure is less than perfect, however, particularly if the technologist has made valuable contributions to the project. Such a policy can lead almost predictably to increasing dissatisfaction with an invisible support role, which is often the lot of the technician/technologist.

Some women technicians/technologists can find a high degree of gender discrimination in a tradition that may permanently restrict their professional growth. It is, in some interpretations, an affirmation of the two-tiered system of responsibility in science—the top tier exercising the authority and consisting almost exclusively of men and the bottom tier carrying out necessary support functions and consisting of most of the women in science. This thesis is presented admirably by Nadya Aisenberg and Mona Harrington[1] in their recent book, *Women of Academe*. The preponderance of female technicians indicates the need for attention to detail and ability to follow complex procedures. But it can also be viewed as an extension of the "quick and nimble fingers" rationale for large-scale employment of women in menial jobs, as in the factories of the early 1900s.

THE RESEARCH ASSISTANT

The third and final category in this artificial pigeonholing of support roles is the research assistant or aide. Here the range of activities gets very broad, but in most instances research assistants are regarded as being near the bottom of the group hierarchy. They do most of the menial chores: washing glassware, entering data into computers, maintaining equipment, doing rough typing, chauffeuring visitors, conducting tours of facilities, collecting field samples—all the necessary but uninspiring work that needs to be done if the project is to maintain momentum. Pay is usually marginal to poor, employment is often short-term, and opportunities for advancement within the group can be scarce.

Since many women occupy niches in this lower realm, it seems worthwhile to examine some of the reasons why they do, and the reasons are many and variable:

- The work environment is usually pleasant and informal.

- Work schedules are often flexible, and part-time jobs are common.

- Work assignments tend to be varied but routine and undemanding, after a short orientation period.

- Research team members are usually congenial, and supervision is supportive if the job is done well.

- Requirements for employment are minimal, rarely extending beyond possession of a high school diploma.

- Institutional benefits—sick leave, vacation time, health insurance—are usually available.

- Close association with enthusiastic and productive professionals can be stimulating at times.

- Opportunities for self-improvement exist, if the job is located on a college or university campus.

The reality seems to be that despite limited horizons and marginal pay, the research assistant job can be attractive, especially to women who have very limited scientific backgrounds and/or extensive family responsibilities. It could seem desirable for women who need a job to provide additional income and want to be treated as individuals participating in a meaningful small-group venture, not as anonymous components of a mass production operation.

Other kinds of research assistants exist, although their job descriptions may not include that title. What about the person with writing or editing abilities who works intimately with the project director, or the one with public relations competencies who handles contacts with the outside world, or the one with financial acuity who maintains internal overview of project spending? These people may not have professional backgrounds in science, but they provide necessary expertise important to a successful scientific enterprise. Assistants with on-the-job experience in these peripheral areas may form part of the long-term core of people clustered around good scientists, providing, along with the technicians, stability for the group in the presence of a continuing flux of postdocs and graduate students.

WHY SO MANY WOMEN?

The reasons for a preponderance of women in scientific support roles are predictably complex, extending far beyond the intuitive one that women have traditionally occupied such jobs. The support positions do require less rigorous training than the professional ones; the support positions have much lower wage scales and are often less than full time; the support positions are more numerous by far than the professional ones; and the support positions often provide much less ego satisfaction. Somewhere in this mix are the selective factors that result in the disproportionately high representation of women.

One selective factor that *was* examined in our survey was a possible gender bias on the part of the project directors/principal investigators in hiring support staff members. No overwhelming conclusions emerged, except that some women professionals admitted that they felt more comfortable with and were more likely to select other women as parts of their research groups, viewing them as potential friends as well as employees. Other women professionals—a majority—emphatically denied any such bias. Most male professionals denied any selection of support staff based on gender (which may or may not be true), but they pointed out the obvious, that most of the applicants for technician and research assistant jobs were women.

We have the distinct feeling that our exploration of the persistent phenomenon of many more females than males in scientific support jobs (other than as research associates) was less than adequate. Is it an illustration of the societal role of women as supportive of male power figures? Is it because women are willing to accept lower-status positions than men? Is it because women are still excluded for many reasons from the training that would place more of them in power positions? We're not sure, but it is possible that the answers to all these questions may be yes.

The reality, though, is that all these women in technician and assistant roles are *in* science, but they are not *of* science in the sense of being full partners sharing in many of the rewards and joys of the profession.

SUMMARY

Most of the support positions in science are held by women. They may be variously classified as research associates, technicians/technologists, or research assistants, and their educational credentials may range from the doctorate down to the high school diploma. Research associates, often with advanced degrees, bring specialized skills to a research group; technicians/technologists, often trained in specific techniques, do most of the necessary routine experimental work and data collection, while research assistants do all the less glamorous chores that permit scientific work to proceed.

Research associate positions are integral components of the structure of modern scientific research, but they can also be traps for women scientists, since salaries are low, job security is minimal, and professional advancement is limited. Probably the best things that can be said for research associate positions are that they provide opportunity for internship with leaders in a specialty area and that they can provide a temporary haven when professional jobs are scarce.

The support jobs—technicians/technologists and research assistant—are vital to the well-being of an effective research team and are usually filled by women. Their activities emphasize data acquisition; their job satisfactions are more in the areas of salary and job security, rather than in scientific recognition.

Reasons given for the numerical superiority of women in support roles range from fulfilling a societal norm (women supporting male authority figures) to needing a job that often provides flexible hours in a pleasant, stimulating work environment. Whatever the motivation, the actuality seems to be that much of the day-to-day work of science is done by women!

REFERENCES

1. Nadya Aisenberg and Mona Harrington, *Women of Academe: Outsiders in the Sacred Grove*. (University of Massachusetts Press, Amherst, MA, 1988), pp. 4–5.

CHAPTER 7

WOMEN SCIENTISTS IN POSITIONS OF POWER AND INFLUENCE

Women as managers of research groups; women as science bureaucrats; women in positions of influence—journal editors, grant review panel members, and technical outreach specialists; why so few women?

Sandra Harding, in her recent book *The Science Question in Feminism*,[1] pointed out the long-standing and still prevalent male perception that women *do* scientific work, but do not usually *direct* it. The perception has been assaulted with increasing frequency lately as more and more scientifically trained women become project leaders, program managers, mid-level and upper-level bureaucrats, and executives in science-related companies. Such positions clearly carry significant responsibilities and power. Decisions about funding for research projects, about recruiting, about promotions and tenure, all are now being made by *females*, and not only by male managers and administrators.

This trend helps to ensure full availability of talent, not obscured by gender. It also gives cause for male unease, sometimes expressed gently by those with conservative attitudes, such as, "Are they competent, or is this some form of affirmative action?" Exploration of whether such unease is justified or not is one of the objectives of this chapter on the acquisition and use of power and influence. Another objective is to examine male and female approaches to the solution of science management problems, to see if different value systems and responses exist.

PERCEPTIONS

Common perceptions of women scientists in positions of power, which they should be aware of, include:

- Women are often tokens of policies dictating the appointment of females at various supervisory/managerial levels regardless of competence.

- Managerial techniques used by women are often flawed by excessive concern for personal needs of employees and by inability to make hard decisions with negative impacts on staff members.

- Female managers are more likely to avoid confrontational situations, or conversely to react harshly and irrationally to them.

- Female managers lose the "invisible woman status," but only for as long as they hold power.

WOMEN AS MANAGERS OF RESEARCH GROUPS

It is certainly a truism that positions of power and influence in science have been occupied traditionally by males. Gender differences in access to training, encouragement, and financial support all have artificially limited women's access to those levels. Correction of this imbalance has only recently been addressed, but even now over 60 percent of the women in science have fewer than 10 years of work experience, as compared with 20 percent of the men in science.[2] This means that the size of the female applicant pool for more senior positions of responsibility is small. In addition, the few women available for such positions are rarely selected for key supervisory, management, or bureaucratic jobs, probably because the final selection process is controlled almost exclusively by men. Also, advancement to levels within reach of such roles is slower for women scientists; this is frequently reflected in their job descriptions and their salaries. The consequence is that women are still drastically underrepresented in key administrative positions.

Women aspiring to management positions in science frequently encounter what has been described as a "glass ceiling," blocking them from reaching the choice top-level jobs. Some analysts point out that whereas overt barriers have largely been eliminated, the more subtle ones persist.[3] These can take several forms, which can be classified as "stereotypes" (women are too emotional to be managers), "unsubstantiated presumptions" (a woman will let family responsibilities interfere with her job), and "obsolete impressions" (when I envision an engineer, I see a man). Such perceptions do not prevent women from entering the work force, but they certainly can retard or prevent women from moving into managerial levels of science.

Women often have support to join the ranks, but the support does not extend to administrative functions.

The prestigious radiobiology department of a Southwestern university lost its chairperson abruptly and unexpectedly. The administration asked members of the faculty to select among itself an interim chairperson to serve until a permanent replacement could be found. Six of the twelve radiobiology faculty members were of sufficient stature to merit the responsibility and coveted power. Of the six, there were two husband-and-wife teams. The two husbands were both known to be highly supportive of women in science in general and specifically had a history of rallying around their wives when unfairly treated (e.g., promotions, invitations to be key speakers, etc.). Curiously, when replacement for the department chair was discussed, neither well-qualified female scientist was mentioned as a possible contender.

Both females admitted to each other following the meeting that they were interested, but quickly realized they did not have the backing of the others, even their husbands. In discussing this enigma some days later, one woman scientist claimed she came to realize that her husband definitely wanted the position and didn't want to compromise his attempt with her nomination. The other woman scientist claimed that her husband definitely didn't want the position, but also didn't want her to have the position because she was already "terminally overcommitted" and spent far too little time at home.

In our survey, women and men both claimed that their managerial strengths were in recognizing talent and motivating talented people. Both genders stated that their greatest weakness was in not being demanding enough. This may be a consequence of scientific backgrounds; scientists *expect* peak performance from colleagues, but they rarely *demand* it. This failure to make and to implement demands was identified commonly as a problem for science managers, both male and female.

Gender differences became apparent in the responses of some scientific supervisors/managers to confrontations. In any laboratory or field environment populated with bright, strong-willed individuals, each with creative ideas they wish to execute, it is not unusual for conflicts and confrontations to develop. The management literature points out that male supervisors/managers tend to greet confrontation situations abruptly and curtly, and the source of dissatisfaction is identified early. Women, on the other hand, state that they are more likely to internalize conflicts and may not take advantage of appropriate early response in the form of comment or constructive criticism. The result can be an unresolved "accumulated debt." When it is finally addressed, it is usually the sum total of the irritation that emerges, often drastically out of proportion to the immediate situation. This invites claims of the "irrational female." Fearing to be misunderstood by either the staff member or the first-line supervisor, a woman manager may prefer written warnings and exchanges of memos instead of direct person-to-person resolution of conflict. This can be considered diplomatic or political, but much time and energy of talented people may be absorbed in the process.

Our survey tried to explore the preferences of male and female scientists for male or female supervisors/managers. Most respondents of both genders claimed that they had no preferences, that they had experienced good and poor of each gender, but on further probing it seems that female scientists tended to be much more tolerant of inadequate female supervision than did male scientists. Counterpoint: Are males more tolerant of incompetent men? Additionally, some female scientists, especially those at entry levels, indicated that they felt more comfortable about approaching female supervisors for travel funds, institutional research funds, and other benefits of their positions.

WOMEN AS SCIENCE BUREAUCRATS

One trend that has been seen in science-related government agencies in the 1970s and 1980s has been the increasing presence of women in mid-level bureaucratic positions, some with substantial power and many with influence on the nature and funding of research projects. A scientific background is usual among the individuals in such positions, but their expertise extends far beyond science, into many social and even political areas; they can become consummate diplomats and astute compromisers, and their networking can be superb. This acquired expertise often becomes a principal component of success in the job, rather than mere scientific competence.

To the perceptive observer, the performance of this rare subspecies—a highly skilled and experienced scientific bureaucrat—can be as rewarding as listening to classical music or attending a good play. There can be great pleasure in being in the company of such professionals and understanding, however dimly, the complexity of their activities— in planning and evaluating, in integrating and guiding diverse groups, in exploring ways to deploy funds, and, most importantly, in gentle manipulation of people, both inside and outside the organization.

On-the-job performances of exceptional female science bureaucrats are even more satisfying to observe, since they are competing successfully in what is still largely a male enterprise. Success factors, in addition to expected high levels of competence and analytic skills, include clearer recognition of all the interpersonal aspects of doing scientific research, greater perceptions of the consequences of actions taken or not taken, evident enthusiasm for projects with which they are associated, and expressed interest in and concern for scientists as people as well as grant applicants. Responses to our questions about female science managers and bureaucrats frequently included very strong endorsements of outstanding examples of women scientists in these roles, especially in funding agencies such as the National Science Foundation.

Women scientists are clearly underrepresented in higher-level academic positions, and a recent study[4] indicates that the same is true for key federal government science and technology jobs. Among the most demanding of those jobs, as identified by the private Washington-based Council for Excellence in Government, only 7.5 percent were held by women. According to the author of the study, women professionals, in competing for top jobs, encounter a "glass ceiling" or an "invisible ceiling" consisting of traditional stereotypes: that they do not have the necessary knowledge and experience, that they are not willing to commit the long hours required, and that their presence may antagonize constituents in technical fields who expect to interact with a man.

The few women in those key science executive positions pointed to the need for increased visibility and ever-broadening contacts if other women are to be included on an unwritten list from which candidates are selected. They also encouraged deliberate involvement in confidence-building public contacts and professional presentations—in activities that ensure presence in a viable position to take advantage of opportunities when they appear.

It must be admitted, though, that some women with scientific backgrounds who are in mid- and upper-level bureaucratic positions in government agencies are there because of equal employment policies pursued from the late 1960s on. Some of them were plucked from entry-level research positions and plunged into the expanding bureaucracies of science-dependent or science-related agencies such as EPA or NSF. Some succeeded, either because they grew rapidly in the job, acquiring the necessary nontechnical skills, or because of some source of continuing political support. They now form part of an important link between academic science and the executive agencies of the federal government. (Undoubtedly it should be pointed out here that many *male* bureaucrats in science agencies were also appointed for political or other wrong reasons, and, as with women, some acquired the needed skills and succeeded.)

Of course, not all bureaucratic appointments work out perfectly. Here is one that didn't:

> Dr. Sandra Spur was a major administrator of a federal agency, hired as a two-year rotator. Before being hired, she demanded that her position be elevated above two career administrators in that agency. Since the agency desired to hire a female, her wish was granted. Her tenure at that agency, however, was short and considered by both males and females quite ineffective. She spent many of her in-service hours surveying the job market. Eventually, she left the agency to continue with her research at a new home base. She was criticized by both males and females for squatting in a position without advancing the discipline within the agency power structure.

Some achieving women in science and science management and other fields admit to occasional bouts with the so-called "impostor syndrome," usually defined as a feeling that they are frauds, that they lack sufficient ability, confidence, and self-esteem to make it in the career world. The syndrome may be caused in part by early imprinting and continuing reinforcement of feelings of inadequacy imposed by the system—*they may have been treated as frauds*. The condition is endemic in the business world where success is characterized by power, domination, and manipulation, but it can surface in science administrators and bureaucrats as well. Men can feel like frauds, too, but they seem more capable of repressing that awareness, as was pointed out effectively in a recent feature article in the *Washington Post* by Cindy Skrzycki.[5]

WOMEN IN POSITIONS OF INFLUENCE

Thus far, the discussion in this chapter has focused principally on hierarchical power—that derived from occupation of a specific box in an organizational chart. We have not yet confronted the larger but less clearly circumscribed area of *influence*, whose ramifications go far beyond the borders of institutionally mandated power and which can transcend any job description. Science is uniquely adapted to the use of influence because of external sources of support and recognition in an extended community outside the confines of the university or agency. Scientists can and do influence the course of events in their universe without occupying so-called "power positions."

Examples of the uses of influence are abundant. Faculty senates participate in tenure decisions; faculty and industry committees evaluate prospective new candidates for positions; colleagues offer opinions on candidates for awards or prizes; colleagues express views on the relative worth of the research results of others; colleagues make nominations for society offices; panels of peers evaluate grant proposals; journal editors decide the fate of manuscripts—all designed to shape, modify, select, change, or improve actions relevant to the progress of science, without the formality of clearly designated authority.

Informal influence is especially important in academia, where hierarchical authority is less clearly defined than it is in government–industry organizations. Recent studies, discussed by Angela Simeone in her book, *Academic Women: Working Toward Equality*,[6] indicate that:

- ". . . women have less power and influence in academic departments than men. They participate less actively in meetings (possibly a reflection of their lower organizational status)."

- "Women have significantly less perceived participation in decision making and job involvement, and more job-related tensions than men."

- "Women are less likely to be part of and to benefit from informal networks of influence and information. . . ."

- "Women reported having less influence over their job situation, greater difficulty getting ideas across to supervisors, and feel less influential in their superior's decisions. . . ."

Interestingly, some of the women in these studies felt that they could have more influence *if they chose to make the additional effort required*. They recognized their own deliberate lack of initiative in pursuing routes to influence, and they also recognized the consequences of not participating actively in the kind of networking that can lead to strong support by male colleagues.

One female respondent made these comments:

> In the early 1980s I was invited to a meeting in San Diego, California. The agenda of the meeting was not clear, but I was lured because the letter of invitation was addressed to "promising young women." At the opening session the first questions raised to this group of nearly 100 was, "Who in this auditorium has aspirations of being the director of NOAA? NSF? NIH? NASA?" No one raised her hand or responded in any verbal manner. In fact, everyone seemed a bit puzzled by the questions. We were then criticized because we were unwilling to set our aspirations at high levels. At the social hour following the session, I found refuge with other familiar women scientists. Our conversations centered on the discussion of the morning. My curiosity was piqued, so I asked: "Did anyone anticipate these kinds of questions?" "Did anyone ever consider being in charge of a national agency?" The response was a clear "no." But the reason for the response should also be noted. The sentiment was "How can I dream of becoming a leader in the national arena when I am neither recognized nor encouraged in my home institution?" "Leadership seems completely out of reach." The resolve of the meeting was that we as "promising young scientists" *should* expand our aspirations. Perhaps a few of us did.

In this section we will explore just a few additional positions of influence: (1) the journal editor, (2) the grant review panel member, and (3) those scientists who translate technical information into language understandable to politicians, lawyers, and the general public.

The Journal Editor

Publication of research findings in reputable journals is an integral part of present-day science. Final decisions about the publishability of papers submitted are made by journal editors, aided by advice and comments from referees and reviewers. Editors want papers that contain good science presented effectively—a not-unreasonable expectation from professionals.

Gender implications in publishing research findings have been identified in the current literature. For example, Sandra Harding, in her recent book, *The Science Question in Feminism*,[1] refers to studies that show that ". . . scientific work known to have been done by women is invisible to men (and to many women) even when it is objectively indistinguishable from men's work, and that a subconscious masculine resistance exists to citing a woman's scientific work." These observations are presented by Harding as obstacles to women's accumulating status within science, and are, in our estimation, overstated; good science properly presented is never invisible, and its citation by peers and colleagues is independent of the gender of the author. While there may be some evidence for reduced visibility as indicated by evidence of relative numbers of citations, none of our interviews provided the slightest evidence for invisibility of women's papers or reluctance of men to cite them, and no journal editor of our acquaintance considers gender to be relevant in any way to the publishing process.

There is another aspect here though—that of women as *editors* of scientific journals. They exist, although not yet in overwhelming numbers. Their appointments usually result from demonstrated competence and expressed interest in the job, or (as with some male editors) from arm-twisting by the executive board of the society publishing the journal. Journal editors are clearly influential and contribute to a significant component of modern scientific practice, but the job is exacting and exhausting, offering truly minuscule rewards for the investments made. The few women editors available for comment saw several positive aspects to their involvement, principally the enhanced visibility, the automatic entry into extensive networks of colleagues, and the opportunity to make a difference in the quality of publications in a specialty area. They saw no evidence of gender biases by the reviewers; only the ability to do the job well seemed to be critical.

The Grant Review Panel Member

We identify in Chapter 10 the "old boys' clubs" in science, pointing out that one of their self-imposed functions is to exert control over grant review panels and to therey guide the flow of research funding. We also suggest that membership on review panels tends to be self-perpetuating in that the same faces (usually male) reappear constantly or are replaced over time by new faces who turn out to belong to students/protégés of the former members. Now a new element has been added; women scientists are appearing in gradually increasing numbers as members of those review and evaluation panels. The easy camaraderie of the club has been disrupted, as its members learn new game rules, new kinds of body language, and new attitudes inserted in the proceedings by female participants. For the able women scientist, membership in these groups presents great new opportunities to influence the direction that research will take and even the selection of those who will be in leadership roles, because of advice that can affect funding decisions about grant applications.

It might be useful to ask a few questions about these relatively new positions of influence that women have assumed. Has the system responded by changing or by resisting change? Are female panel members really listened to by

the male majority? Do women scientists often assume assertive roles in the discussions? Do the agency participants and
grant monitors take comments by women members seriously? Answers to these questions show, as usual, high
individual variability. Some women panel members still feel
reluctant to dominate or even to participate in extended and
controversial discussions. They feel that their comments are
more readily discounted by male chairpersons and male
panel members. Many more women feel, however, that
scientific credibility and *assertiveness* are the two keys to effectiveness in panel discussions; that regardless of gender, if a
person is known to be a productive and informed contributor in a specialty area and talks forcefully from that power
base, he or she will be listened to in matters related to that
area of expertise.

Women scientists already know or learn quickly the
many devices that increase effectiveness in group discussions—such as maintaining a persistently positive civil attitude, providing support for relevant opinions offered by
other panel members, refusing to be interrupted or "talked
over" by other members, insisting that critical consensus
points be recorded by the rapporteur, and providing the
proper visual cues indicating their knowledge of prevailing
(largely male-oriented) game rules.

As an example of how women scientists engage effectively in discussions when surrounded by male panel members, we offer our observations on the performance of Dr.
Nancy Rittgers at a recent executive session during a two-
day Sea Grant site visit.

> Dr. Rittgers is an authority on invertebrate biochemistry and
> has been a participant in many review and evaluation panel
> sessions. She chose a "power position," seated opposite the
> chairperson at the usual open square table arrangement, and

placed notes and folders of reference materials in front of her, to establish physical "turf." She quickly introduced herself to panel members seated adjacent to her and carried on polite conversation with them until the meeting began. During the session her comments established credibility and indicated careful scrutiny of the proposals to be considered by the group. When relevant, she offered clear decisive opinions and firm responses to rebuttals. At several points in the discussions she was interrupted rather peremptorily by other (male, of course) panel members; she handled these intrusions smoothly but firmly and went on to make her points. She presented her own reasoned analyses of issues at appropriate moments in deliberations. At the conclusion of the session it was evident from the chairperson's summary that her contributions had made a significant impact on the thinking of the group and on the nature of the advice to be given the funding agency.

An initial reaction to this example might be "This is the way that *any* effective panel member *should* perform"—and that is precisely the point that we would make here. *Gender* is not crucial in today's climate, but a professional approach to interpersonal exchange *is*; that approach requires a combination of credibility, assertiveness, and knowledge of networking game rules. Women scientists who tend to be intimidated by the often intense atmosphere created by group discussions can easily adopt many of the behavior patterns exhibited by "dominant males" and quickly move into positions of greater influence in the proceedings. The alternative is to change the rules.

An example of the influence of "a woman's voice" was described in one of our interviews:

Dr. Gloria Towe was the sole female on a committee of many at a federal agency. The task at hand was to decide the four winners of a prestigious award. There were many candidates and multiple meetings. Eventually, the "top ten" candidates were rank ordered. Each was male. Formal and informal discussions

ensued to advocate no. 5, no. 7, and no. 10 over no. 1, no. 2, and
no. 3 before the final round of voting. The rationale in each case
centered on the importance of the award to the advancement of
the careers of nos. 5, 7, and 10.

At this point, Dr. Towe initiated the advocacy for females
who fell below the top-ten threshold but within the top twenty.
Her plea included merit plus importance of contribution(s) to the
scientific discipline, plus the significance of the award to the
careers of named persons. An opinion was voiced that awards
are more important for men than for women based on men's *need*
to succeed. The opinion was initially defended, but eventually
silenced. A motion was made and seconded to include the top
twenty in the final round of voting. The final results: awards
were presented to two men and two women. Dr. Towe credits the
turnaround to the fact that she held back comment until the
critical time.

The Technical Outreach Specialist

Much of the scientific research conducted now has
immediate application to issues and problems currently
confronting the human species—environmental degrada-
tion and its consequences, public health, food supply, energy sources, and armed conflict. The direction and intensity of research in any area are determined by funding
levels, which make scientists vulnerable to political processes, which are in turn shaped to some extent by the
perceived needs of society. Public perceptions therefore
become of significance, and scientists have been remarkably
ineffective in communicating the importance of their research findings to others outside their own severely restricted collegial spheres.

This deficiency has been recognized and bemoaned, and
fortunately a small segment of the scientific community has
developed what are referred to by sociologists as "outreach

skills"—abilities to translate and synthesize results of research into packages that are intelligible to nonscientists (politicians, resource and environmental managers, lawyers, and the general public). We need more of these rare individuals, if only to avoid the misrepresentations and distortions of science that are commonly found in reports prepared by the news media.

People who inhabit this critical zone—part scientist and part public relations expert—appear in television documentaries, are quoted in feature stories, and participate in various lecture circuits. They may or may not do active research, but they have the competence to synthesize information in a specialty area and to do it interestingly and comprehensively.

Professional women are appearing in greater numbers in these translator/synthesizer roles, not as public relations aides, but as *authorities* in their fields of research interests. Proportional numbers are still small when compared to men; this is likely a reflection of the gender composition of the total scientific population. Of the women scientists interviewed by us, roughly one-fifth indicated that they participated in some form of outreach activity. Channels included contributing on-camera segments to television documentaries, lecturing on the service club (Rotary, Lions, Chamber of Commerce, Exchange) circuit, lecturing in continuing education courses for adults, writing articles for semipopular magazines or periodicals, assisting media feature writers in preparing interpretive stories, working with legislative aides in preparing background documents supporting draft legislation, and presenting seminars on specialty areas to other professional groups (physicians, lawyers). The point to be made here seems to be that, with a larger pool of women scientists to draw from, a larger representation is being observed in public contact and interpretive roles that are a necessary part of the current practice of science.

SUGGESTED ACTIONS

Of critical importance in any managerial position, scientific or otherwise, is *analytic competence*—the ability to examine a mass of information and to *synthesize* this into the big picture relevant to planning and implementing a course of action. Women in power positions must be imaginative, assertive, and even aggressive, but diplomatically so.

Success in group interactions is a primary requisite for the woman manager. Discussions should be led toward areas of personal competence and areas of ignorance should rarely be disclosed voluntarily. The correct and relevant questions should be asked, and done so with ease. Contributions to discussions should be sharply focused and never long rambling dissertations, couched in bureaucratese and replete with current buzzwords. The appropriate uses of diplomacy by women administrators can be effective in group discussions in which men predominate, but it must be as a complement to and not as a replacement for competence. The managerial woman had better be good as well as diplomatic.

Beyond the basic requirements of intelligence and training, women scientists must portray the *image and demeanor* appropriate for administrative/leadership positions. The image is fluid, however, and is changing at the moment. Dressing for managerial success is no longer as severe and formal as it was a decade ago. Feminine clothing is now considered appropriate and desirable, as long as it does not convey the wrong impression. The impression should be professional, not provocative.

Women managers/administrators have a real need to get up to speed quickly as team players—a difficult added burden since the concept itself has not been a part of the normal early background of females, even in modern society. The problem can be especially severe if previous science experience has been as an independent researcher; it constitutes less of an impediment if previous experience has been as part of a large research group.

SUMMARY

Women scientists are appearing more frequently in positions of power and authority, but their numbers are still proportionally very small. The disparity is in part a result of persistent numerical imbalances that begin in graduate school or earlier, in part because the size of the female applicant pool for senior managerial positions is still small, and in part because the final selection process for positions of power is still almost entirely controlled by men. Despite these impediments, women are assuming larger roles as project leaders, program managers, laboratory directors, science bureaucrats, and executives in science-based companies.

From our study, and those of others, several tentative conclusions may be proposed about women as managers of scientific organizations:

- Female and male science administrators have similar perceptions of their greatest managerial strengths and weaknesses. They find strong abilities in common in recognizing talent and encouraging talented people and common weaknesses in demanding peak performance from all members of their staff.

- Many women in science management positions are there because of demonstrated ability and productivity and are highly selected survivors of a male-dominated system. Some, however, may be in their jobs because of institutional affirmative action. (This, of course, does not necessarily mean that they may not grow in the job; nor does it ignore the fact that many men, too, are selected for a plethora of questionable reasons and may grow into the job.)

- There is some evidence that women bring special qualities to supervisory/management roles in science. In general, they tend to be less confrontational, more people-oriented, and often much more perceptive than their male colleagues.

REFERENCES

1. Sandra Harding, *The Science Question in Feminism* (Cornell University Press, Ithaca, N.Y., 1987), pp. 70–81, 64.
2. National Science Foundation, *Women and Minorities in Science and Engineering* (NSF 88-301, Washington, D.C., 1988).
3. Katherine Naff, Probing the glass ceiling, *Commerce People*, (June 1991), p. 10.
4. Marcia Clemmet, Toughest federal science jobs elude women, *The Scientist* 4(20), 8 (1990).
5. Cindy Skrzycki, Healing the wounds of success, *Washington Post* (23 July 1989), pp. H1 and H3.
6. Angela Simeone, *Academic Women: Working Toward Equality* (Bergin and Garvey, South Hadley, MA, 1987), pp. 87–90.

CHAPTER 8

WOMEN SCIENTISTS AS MENTORS AND ROLE MODELS

Duties and responsibilities of mentors; intrusions of personal considerations into mentor–protégé relationships; the value and scarcity of female role models; the negative role model.

Terms that recur frequently in our discussions and in responses to our questionnaires are "mentor" and (especially among women respondents) "role model." Mentors, in the best and broadest sense of the word, can be critically essential to the career development of protégés. In the words of Niki Scott in her syndicated column, *Working Woman*[1]: "Mentors hold the signposts for us on our journeys toward competence, assertiveness, and self-esteem. . . ." They assume responsibility for imprinting the essential components of science—its history, traditions, ethics, value systems, approaches to thinking, analytical methods, evaluation criteria,

145

publication guidelines—all the factors that are important to the continuity of science and the achievement of success in a chosen field.

MENTORS

Mentors can serve variously as advisors, confidants, parent figures, confessors, inspirational sources, door openers, motivators, judges, role models, friends, and of course teachers. They do not, however, have to fill all of these roles, since each mentor–protégé relationship will be slightly different, depending on mutually perceived needs and on the personalities of the two participants in the joint venture.

Evolution of the mentor–protégé relationship, especially as it applies to business activities, was explored extensively by Leslie Westoff in her recent book, *Corporate Romance*.[2] She reminds us that the original Mentor, in Greek mythology, was a wise and trusted advisor, but that the mentor concept expanded in the Middle Ages to encompass the master–apprentice relationship in which total responsibility for the moral and technical development of the apprentice was invested in the master. More recently, as Westoff views it, the mentor's responsibilities have become more narrowly defined (at least in business relationships) and can be more accurately described as "sponsorship," designed to ensure that the protégé will move up in the system, and in which much of the educational component of the relationship has been subsumed by other experiences. It seems that in science we have retained more of the classical connotation of the mentor as advisor, but with a liberal sprinkling of sponsorship laid on as well.

During the interviews that we conducted, the contributions to successful careers that have been made by mentors were emphasized repeatedly. Examination of their functions that were considered to be of particular importance resulted in the following list:

- Demonstrating a style and methodology of doing research

- Developing an analytical approach to selection of significant questions and to choosing appropriate approaches to solving them

- Discussing the concepts in any subdiscipline, and the evolution of those concepts over time

- Exploring and evaluating the literature of the discipline and the broader body of knowledge of which it is a part

- Discussing the ethical basis for scientific research

- Considering, analyzing, and evaluating the work and conclusions of colleagues

- Transmitting, by example and discussions, the skills required for effective scientific writing

- Evaluating successful teaching techniques

- Facilitating access to the research community in the discipline (scientific societies, peer groups, international science, "in groups," etc.)

- Illustrating the methodology and significance of "networking" in science

- Developing attitudes and approaches to the many

interpersonal relationships involved in being a scientist

Obviously, not every mentor–protégé relationship will be the perfect embodiment of all these functions; but one of the most satisfying aspects of our study was the frequency with which good mentors were identified and applauded for their efforts—especially when all the other demands on professional time were considered.

ROLE MODELS

Role models are also important in career development. A mentor may be a role model as well, but role models need not be mentors; in fact, no personal relationship may be involved. Perceptive young scientists often recognize and try to emulate professional behavior patterns of exceptional scientists, or they may aggregate and attempt to integrate the favorable characteristics of several such models. This too is part of imprinting—transmitting ways of thinking and acting that distinguish the outstanding professional from the amateur.

Even in instances where incipient scientists have not been fortunate enough to participate in a good mentor–protégé relationship, it is rare that role models have not had favorable impacts on emerging careers. Listening to a superb presentation at a scientific meeting, or attending a brilliant classroom lecture by a respected faculty member, or reading an exquisite journal article, or even observing the conduct of a socially conscious professional at a cocktail party—all offer insights for the discerning novice. Role models may know that they are good, but they may not always be aware that they are being scrutinized closely for traits useful to aspiring professionals. Scientists, like most humans, are remarkably reticent about complimenting those from whom they learn the most.

A good example of the reality of "unconscious role modeling" was recounted to us recently during one of our interviews.

> A colleague, Dr. Joanna Ellerby, with a long list of excellent publications, was approached by a graduate student representative from a distant university with a request that she present an invited lecture on scientific writing, which was to be part of the ceremonies honoring a retiring faculty member in the English department. Apparently Dr. Ellerby's published papers had been used extensively as outstanding models of scientific prose in a graduate course on technical writing given for many years by the retiring professor. Dr. Ellerby was totally unaware of this distinction and was more than mildly flattered that her publications had such utility, beyond their scientific value.

This discussion of mentors and role models has so far been remarkably gender-free, and logically it should be. But there are aspects of these roles that do display some variability when examined from that perspective. Some common perceptions that may or may not be true are:

- Female graduate students will deliberately choose female faculty members as advisors and mentors, if other selective factors are not overriding.

- Mentors who are female develop and encourage a more personal relationship with female protégées than with males.

- Mentor relationships are more important to female scientists than to male scientists during their early career progressions because of a long tradition of female dependency.

- Some of the success of women scientists may be linked to long-term professional relationships with one or more male mentors who are recognized leaders in a specialty area.

- The concept of women acting as mentors and role models for other women is relatively new, with its principal emergence coinciding with the success of the feminist movement of the past few decades and the gradually increasing assumption by women of more senior positions in science.

Disregarding, for the moment, the extent of validity of these perceptions, the male–female components of a mentor–protégé relationship can't be ignored. Even though the association is a professional one, it is entered into by *people*, all of whom bring their own baggage of emotions, attitudes, prejudices, hang-ups, and needs. Male mentors of female graduate students and junior scientists must recognize and accept additional guidelines and strictures—beyond those that would apply to male protégés—to ensure that any aura or suspicion of sexual involvement is dispelled, since the most straightforward and businesslike of arrangements may be misunderstood or misinterpreted by colleagues and casual observers.

It is true, though, that the kind of long-term contact required in a mentor–protégé relationship can and sometimes does lead to very personal interactions—to affairs—if the female protégée has a tendency to fall in love with men she admires or if the male mentor fails, unconsciously or deliberately, to maintain some professional distance in the relationship.

Males, too, have a tendency to fall in love with a female mentor. It seems that the transformations of the mentor–protégé alliances into personal and even physical relationships are not uncommon in science; in fact, our data suggest that they are remarkably prevalent!

Leslie Westoff in her book *Corporate Romance*,[2] referred to earlier, outlines excellent advice for male mentors and female protégées in a chapter titled "The Mentor Trap." Although oriented toward relationships in business situations, much of her discussion is relevant to science mentor–protégé associations as well. For the mentor (read middle-aged male professor) confronted by a prospective protégée, some tidbits to remember (as extracted from Westoff's discourse) include:

- "Don't discuss personal matters or give personal advice."

- "Don't compromise personal values for a superficial relationship."

- "Try to find the right combination of objectivity and friendliness."

- "Always keep the office door open during conferences."

- "Avoid the evolution of a 'father/daughter' relationship, especially one characterized by undue deference displayed by the protégée."

- "At the first sign of personal attraction on either side, dissolve the relationship quickly and completely."

- "By allowing a romance to happen, the mentor takes unfair advantage of the protégée, who may easily become infatuated with the glow of power and influence."

We might add to this listing the admonition that university administrators are very sensitive about the twin dangers of legal action connected with sexual harassment claims by students and the consequent negative public image created for the institution. Some universities have taken vigorous actions to break up or discourage faculty–student liaisons. Laura Mansnerus, in a recent article in the *New York Times*,[3] describes policies at Harvard and the University of Iowa that prohibit "amorous relationships" between faculty members and anyone under their supervision. Certain other universities, according to the article, have taken a different approach, dictating that "if the student who was sexually involved with a teacher later brings a sexual harassment complaint, *it will be presumed to be valid.*"

For the female protégée entering a professional relationship with a male mentor, a distillation of points made by Ms. Westoff[2] includes:

- "Be more assertive and less deferential in developing the relationship."

- "Men are reluctant to be mentors to women, principally because of concern about how others will view the relationship."

- "When a woman is attractive, she finds herself with many offers of 'professional friendship' that turn out to be the same old sexual game in a new setting. . ."

- ". . .there are many similarities between a growing mentor relationship and the evolution of a romantic intimate relationship."

- "A sexual relationship with a mentor is almost certain to have a negative effect on any future career."

- "If a sexual relationship develops, make sure that emotional boundaries are clearly defined."

Westoff offers other suggestions that have relevance to scientific mentoring. Those with the greatest potential for academic application are to *institutionalize* the mentor system by randomly assigning all senior staff members to act as mentors for *junior staff* (as well as for graduate students), to develop training programs on how to be a mentor, and to make successful mentoring an element of performance evaluations for promotions (and even for tenure). Westoff further stresses that "The mentoring system should not be left to chance or to develop randomly; its nuances and virtues should be taught and discussed, as should effective management of the mentor–protégé relationship."

But what if a too-personal—even sexual—relationship should emerge from a mentor–protégée association, despite all admonitions, and what if the relationship becomes common knowledge? Here the potential for serious damage to both partners exists, and additional strictures apply:

- Middle-aged faculty members are part of an actuarial group at particularly high risk from the perils of poorly managed sexual adventures with graduate students.

- Most university administrators hold very conservative views about proper faculty–student interactions and may react strongly to transgressions. The erring faculty member may be warned or censured and the wayward student may be invited to leave the department or the campus permanently.

- Faculty colleagues may find private glee but may

express public dismay at an obvious lack of good judgment and moral standards on the part of a presumed professional.

- Examples of this nature can serve as warnings to other male faculty members to permit only the most formal interactions with graduate students of the opposite gender, thereby depriving many female students of the advantages of informal mentor relationships.

- Even the suspicion of a too-close mentor–protégée relationship can put the student at a disadvantage with other faculty members, who will look for evidence of favoritism and special treatment. Furthermore, that student will be assumed to lack commitment to the discipline and rigor of scientific training and to be looking for a too-easy entry into the community of scholars. Other faculty members may close ranks against the student in an attempt to ensure that undue advantages do not accrue from the relationship.

- Other graduate students may react negatively by excluding the miscreant student for proving an unfortunate stereotype.

It seems from all this discussion that perhaps we need to develop an immutable code of conduct governing mentor–protégé interactions in science comparable to those already prescribed for lawyer–client and doctor–patient relationships, universally accepted and usually adhered to as part of standard professional ethics. A not-insignificant part of that code would focus on managing the sexual attractions that may emerge unexpectedly.

While most of this discussion may imply that the male

is the mentor and the female is the protégée, there are many cases where the reverse is true. With increased numbers of female faculty and expanded power, the proportion of female mentors and male protégés is likely to increase. The associated concerns would be no different.

Such intrusions of emotions and sexuality into professional relationships should not, however, denigrate the importance of the mentor role in career development. Gender should be identified as a complicating factor, but not an overriding one, if the parties involved retain a genuinely professional outlook.

Some realities of the mentor–protégé association, as deduced from our questionnaire responses, are these:

- There is a tendency for beginning graduate students who are female to gravitate toward female faculty advisors, unless there are compelling reasons related to specialty areas to select a male advisor.

- There is little evidence for the existence of a counterpart tendency on the part of beginning male graduate students to gravitate toward male advisors. Here the choices seem to be primarily discipline-oriented, regardless of the gender of the potential advisor. There is also the reality that the proportion of female faculty is smaller.

- Male faculty members almost unanimously deny any gender considerations in accepting graduate students, as well they might in today's climate favoring reduction of subtle discriminatory practices.

- Similarly, most female faculty members deny any gender-based selection, but many admit that their graduate students are mostly female.

- The significant role of a good mentor, regardless of gender, in career development has been reaffirmed repeatedly and often emphatically by male and female respondents alike.

- Many women scientists affirm the importance of one or more male mentors at various points in their careers, pointing often to the critical support functions of penetrating male-dominated networks.

- Mentors, of course, have additional functions such as encouraging performance that is consistently above expectations, applauding assertive behavior, building feelings of self-worth and excellence, and advising in periods of critical career decisions.

It seems reasonable to conclude that the role of mentor, while it may have some superficial gender distinctions, is clearly one in which women scientists achieve equal recognition for equal efforts, but that there are problems with numbers. An observation made repeatedly by younger female respondents in our study was that there were numerically so few senior women on the faculty in their discipline with whom graduate students could relate. The concept of women as mentors and role models diminishes in value with physical distance, too, so that identification with a successful woman scientist in another part of the country is difficult. All of this is, of course, a reflection of early discriminatory practices against women and continuing gross imbalances in male:female faculty ratios. The concept of women scientists as mentors and role models is an excellent one; its fulfillment is severely restricted by numerical realities.

Exclusion of women from graduate education in the sciences has diminished remarkably in recent decades, but some of the *effects* of earlier overt exclusionary practices can still be seen in the small percentages of women in more senior faculty positions in the quantitative sciences. As an example, Princeton University did not accept women in graduate physics until 1971 and in graduate mathematics programs until 1976![4]

A new study conducted by the University of Rhode Island for the Oceanographic Community[5] indicates that there is a strong gender bias in the numbers of persons awarded master's degrees who are then encouraged to go on as Ph.D. candidates. Whether this is discrimination, lack of subtle mentor encouragement, or female choice to terminate after a master's degree is a complex issue and is yet to be determined.

One final qualification must be placed on the seemingly obvious notion of senior women scientists as role models for female students. There appears to be no strong evidence, other than anecdotal accounts of individual cases, that the presence of female role models has a significant influence on career choices.[6] The idea that enhanced recruitment of women students may depend on the presence of prominent women scientists as role models seems to be based only on assumptions, and the effect may be statistically only minor and may be even negative if the climate for women does not improve. Despite this apparent lack of quantitative information, however, the number of instances, in our studies and those of others, in which women scientists mention role models suggests that more data are needed before the idea is discarded or discounted.

A more curious suggestion is that many women science professionals might in fact be serving as *negative* role models. Once young women get a close-up view of professional women scientists and detect the degree of overcommitment and frustration at trying to juggle multiple priorities that often override the joys of doing good science, they might well decide, "I don't want to be like this!"

SUMMARY

The mentor relationship has evolved over the centuries; its present embodiment in science is a senior advisor who also acts as a sponsor for promising protégés. Sponsorship includes provision of access to the thinking, attitudes, and the networks of science. The value of the mentor has been reaffirmed often, in our study and in those of others, as critical to the development of new professionals.

The mentor–protégé association is not entirely gender-free. Problems of a too-close personal relationship between a faculty member and a student can destroy the academic advantages and produce severe career impacts. Fear of such entanglements causes many male academics to eschew any but the most formal of interactions with female students, thereby depriving those students of very necessary introductions to the networking and collegiality of science— advantages that then accrue principally to male students.

Mentors often serve as role models for students, although role models may not necessarily be mentors, but merely exceptional scientists whom students would like to emulate. The scarcity of senior female scientists who can serve as role models for female students has been described repeatedly. Earlier discriminatory practices have been identified as the cause, and continuing female–male numerical imbalance in university faculties perpetuates the problem. One ancillary observation that has some relevance is that some overstressed or hostile women faculty members may actually serve as *negative* role models—as examples of what a career should *not* be.

Despite such occasional limitations, the mentor and the role model can provide significant contributions to the professional development of young scientists. Intrusion of gender is unfortunate but still a reality. Closer adherence to a code of conduct to reduce personal components of mixed-gender mentor–protégé relationships might prove effective, as could greater insistence by female graduate students on participating in all of the collegial components of the relationship.

REFERENCES

1. Niki Scott, Show your mentor the gratitude you feel, *The Star Democrat* (Easton, MD) (21 March 1990), p. 3C.
2. Leslie A. Westoff, *Corporate Romance* (Times Books, New York, 1985), pp. 123–151.
3. Laura Mansnerus, Colleges break up dangerous liaisons, *New York Times, Section 4A Education Life*, (7 April 1991), pp. 1–2.
4. K. C. Cole, Who needs women? *Omni* 9(8):35 (1987).
5. Margaret Leinin, Women in oceanography at the University of Rhode Island and other oceanographic institutions, discussion at the Oceanographic Society Meeting, February, 1991, St. Petersburg, FL.
6. Jonathan R. Cole, Preface to the Morningside edition, in *Fair Science: Women in the Scientific Community* (Columbia University Press, New York, 1987).

CHAPTER 9

THE MOBILE WOMAN SCIENTIST

*Some reasons for the itinerant behavior of scientists;
mobility as affected by marriage and family—changing
attitudes; special challenges for dual career couples;
the "trailing spouse."*

One outstanding characteristic of scientists is that they are geographically flexible—they move around a lot, from one position to another, especially in the early phases of their careers. Motivations for moving are varied, but some of the more common ones are:

- Dissatisfaction with progress in rank or salary at the current location
- Offer of a comparable or better position at an institution with more prestige
- Transfer from a junior college to a four-year institution

- Transfer from a private college to a research university in the interests of decreased teaching load and increased possibility for scholarly production

- Disenchantment with an entry-level industrial research position and desire to do research in an academic environment (or the reverse of this)

- Negative tenure decision or unavailability of tenure at the current location

- Perceived absence of opportunities for promotion in the current organization or specialty area

- Move from a nontenured postdoctoral or research associate position to a tenure track faculty position elsewhere

- Disagreement or lack of rapport with department chairperson or dean

- Transfer for personal or family reasons from urban university to rural university, or vice versa

- Divorce or death of spouse, and a strongly felt need to relocate

- Disenchantment with entry-level government position, usually because of overwhelming paperwork and overload of nonscientific chores, followed by a move to an academic position

- Disappearance of funding for a research associate position at the current location, combined with absence of institutional funds

- Move of spouse/partner to a new job in another location

ITINERANT BEHAVIOR

The list could be extended to chapter length without exhausting the possible reasons for moving, but, whatever the motivation, scientists normally do move a number of times during their careers. This is accepted as standard procedure and a way of life for upwardly mobile professionals. Aspects of these moves for career purposes that should concern us in this chapter are those that may be related to gender. Patterns are obviously not going to be the same for male and female scientists who are married if they follow the somewhat dated societal dictum that the male is the principal provider, and the wife and family go where and when he goes. There are, however, clear indications that such a dictum is no longer relevant for substantial numbers of today's women scientists.

Responses to our questionnaires and interviews left no doubt that women scientists (80 percent of them in our case) felt themselves to be mobile, although half of this number made it clear that ties to spouses/partners/families would temper their decisions to move. To us, two very interesting additional findings were that of these mobile women scientists, 25 percent stated that they would be willing, for career reasons, to live separately from spouses/partners and 18 percent stated that their spouses/partners would move if the woman's career so dictated. We have the feeling, though, that these statements need to be probed much more deeply before definitive conclusions can be drawn. Living apart from a spouse/partner may be okay—even great—for a short time, but what about it as a way of life? Also, pious words from husbands about throwing away their jobs and moving when the wife's career dictates should be viewed with some circumspection in the absence of tests of that piety. But even if these percentages are inflated by a tad of wishful thinking on the part of the women scientists we queried, the *concepts* of living separately for career purposes or priority of the woman's career in decisions about moving *exist* and are acceptable to many female professionals.

But what of the other respondents who said in essence, "Yes, I'm mobile, but I'm also tied to a spouse/partner"? That segment of the sample population would probably subscribe to some or all of the following statements:

- Despite modernization of views on marriage and family life, the mobility of a married female scientist is still somewhat dependent on the spouse.

- Living apart is not marriage; it is merely being tied to a partner.

- Occasionally the wife's job, if it is a good and satisfying one, can be an overriding factor, but such instances are still unusual.

- Moves that are dictated by acquisition of a better job for the husband often force the wife into jobs in the new location with less prestige and pay, or into part-time teaching jobs, or temporary "research associate" jobs that may depend on the continuity of someone else's grant. Career progressions can be disrupted by such forced moves.

- Divorce or death of a spouse is a strong motivator for mobility, either for male or female.

EXAMPLES

Examples of the forces operating for or against mobility abound in our interview data. Here are three:

In the early 1970s Drs. Gloria and George Roman explored three position alternatives. Nepotism rules were cited to keep them in separate departments, so that he would not be her boss and she would not be his boss. The maneuvering became so involved that they declined all positions and started a small nonprofit research foundation in a nondesignated location. The Romans chose a community where they desired to reside and raise their family. The laboratory became a refuge for many professional couples with similar scenarios.

Dr. Jan Murphy was a young, enthusiastic, vibrant administrator at a new and emerging state science-related agency. In her four years, she accomplished far more than was on the agenda set by herself and the governor. Her accomplishments were acknowledged by both males and females. Six months before her resignation, she announced that she was taking a job as associate director at a small nonprofit academic institution. In fact, she reasoned, the amount of travel required too much separation from her elementary school-aged child. Her new job had local responsibilities. She was damned by some people for her unwillingness to stick with it, to persevere.

Dr. Jane Bouchart married another scientist about twenty years her senior. They both held enviable positions in different departments at a Southwestern university. They enjoyed raising several children. When Dr. William Bouchart retired, Jane was at a high point in her career. For the first time there was a tug away from her dedication to her profession. William wanted to travel and write; Jane needed to be in her laboratory. They both anguished over their future. "If I push on for the next twenty years before my normal retirement, William, if he is still around, will probably not enjoy the travel we both want." Jane worked for three more years and then gave up her position. She continued to participate on a consulting and informal basis. She is content with her decision.

It seems, though, from our almost endless discussions, that an interesting and viable middle ground is developing for husband–wife teams in science that extends well beyond the stereotype in which the man is the high-salaried sought-after professor and the wife is given the "courtesy" of an unpaid or underpaid research appointment. This middle ground does not require that lives be spent apart or that the woman invariably makes decisions about moving based only on her career priorities. It is best exemplified by a variety of husband–wife combinations that preserve and foster the career interests of *both* partners. Some specific cases that have been reported or that we have observed are:

- The woman is on the university faculty while the man is a partner in an environmental consulting company.

- The man directs a small specialized private research laboratory while the woman is a professor at a nearby private college.

- The woman is a mid-level scientific bureaucrat in a federal granting agency while the man has a counterpart position in a state environmental conservation office.

- The professional couple has agreed to a "split appointment" whereby they would share an office and divide the salary and teaching responsibilities of a single faculty position.

- The man is a leading candidate for a dean's position and the woman has been offered a tenured position at the institution.

- The man is on the research staff of a big city medical school while the woman is on the botany faculty of a small college in the same city.

Problems can emerge if the careers of both partners are too closely intertwined. A recent examination of married couples who are members of the National Academy of Sciences[1] indicated an underlying struggle by the woman member of the pair to achieve and retain a separate scientific identity, rather than being an adjunct to the man. Although the examples cited in the report emphasized individual accomplishments, the difficulty in keeping personal and professional lives in balance was alluded to repeatedly.

To us, one of the most interesting observations Elizabeth Pennisi[1] makes in describing the findings from the NAS study is that the men, looking back on successful careers, stressed that their research had been a labor of love, whereas the women recall struggles against prejudice and exclusion. The women did, however, feel that the situation for women and for two-career couples is improving, and that equal career opportunities are now more accessible than they were previously. Other authors[2] stress that dual-career couples offer a challenge and opportunity for enlightened institutions, with flexibility and innovation the keys.

Critical aspects of these husband–wife combinations are that the career interests of both partners are considered as being *precisely equal in importance*; that the partners' careers and scientific identities are *totally independent*, so that the success of one is unrelated to success of the other and is an individual matter; and that the subset of the scientific community that serves as *colleagues* for one partner is totally different from that encountered by the other. Any discussion of a change in location—should a move seem expedient—becomes an evaluation by two *peers* of all the professional and personal factors important to a decision pro or con. Arrangements like these seem to reflect some advancement in attitudes, and even if some of them degenerate into episodes of recrimination or name-calling during periods of crisis, the *principle* of equal career consideration for both partners is apt to persist and to be strengthened in future decades.

Interesting peripheral problems can be created with the so-called joint appointments. Some academic institutions and most industrial research laboratories still discourage or have stated policies preventing employment of husband–wife teams as members of any single group. Concerns expressed by industries seem to focus on difficulties with supervision and possible compromise of proprietary information and patent rights. When joint university appointments are made, with tenure to both partners as in one of the examples cited above, the result can be a downward displacement in tenure considerations for existing faculty members, with accompanying dissension. Split appointments may not always be so great either. The initial agreement to share one office, one laboratory, one salary, and teaching responsibilities may be short-lived, to be replaced quickly by demands for two salaries, two laboratories, and two of everything else—which can be perceived as a device and can be a source of dissension among other faculty members. In one example of a split appointment cited above, the split actually worked out to be three-way, with a baby occupying a crib in one of the two offices that were eventually assigned to the couple.

One additional insight that emerged from our study has to do with the old matter of perceptions—specifically, the perception that married women *won't* move, so, as a consequence, *they are never even considered or asked to consider positions* by academic, government, or industry recruiters. To some female respondents, such an erroneous perception unfairly excludes them from jobs for which they are well qualified and in which they might be interested. They feel that they are denied access for reasons totally divorced from any scientific rationale.

SUMMARY

Scientists seem to move around from job to job frequently, for a long litany of personal and career reasons. Traditionally, married women scientists have had to assume the added burden of the need to accommodate to changes in job locations of their spouses. In our survey, 80 percent of the women scientists queried described themselves as mobile, although half of these indicated that their mobility would be tempered by ties to spouses and young families.

An emerging phenomenon is the increasing prevalence of dual-career husband–wife teams in science and the development of arrangements that foster the careers of both partners. Such arrangements may include equal consideration when changes in job location are considered, joint or split appointments, and enough built-in flexibility to permit joint child care without impeding the career advancement of either partner. Evidence from several studies, including our own, indicates that the situation for women and for dual-career couples is improving.

Several aspects of the mobility of married women scientists deserve further attention. One is the assertion by 25 percent of those queried in our study that they would be willing, for career reasons, to live apart from spouses. Another is the statement by 18 percent of the respondents that their husbands would move if the wife's career required it. The credibility of such assertions may be questioned, but even if percentages are inaccurate, they indicate fundamental changes in thinking about mobility and professional careers of women.

REFERENCES

1. Elizabeth Pennisi, Two generations of NAS couples reflecting changing role of women, *The Scientist* 4(23), 1 (1990).
2. C. Sue Weiler and Paul H. Yancy, Dual-career couples and science: Opportunities, challenges and strategies, *Oceanography* (November 1989), pp. 28–31, 64.

CHAPTER 10

PARTICIPATING IN THE FRATERNITIES, CLUBS, AND SOCIETIES OF SCIENCE

The "invisible colleges" of science; membership in the "old boys' clubs"; the fraternities and their pledges; visible and invisible networks.

Science is a profession with the principal purpose of acquiring and publishing new information about the physical universe. The practice of science involves many personal interactions, some of which contribute greatly to the joy of "doing good science." Membership and active participation in the social groups that develop among professionals are to many a source of pleasure and professional advantage, even, for some, transcending the usual rewards of promotions and awards. Invitations to membership in these unstructured, informal groups of colleagues—called "invisible colleges" or "clubs" by some authors—are in themselves forms of recognition of accomplishment accorded by scientists to scientists. The groups also serve as important networks for communication about science as well as such weighty matters as job openings and impending publications. They constitute major bonding mechanisms among good scientists in particular specialty areas.

In our questionnaire we asked "How do you relate to the 'in-groups' or 'scientific clubs' in your discipline?" It is notable that not one woman in our survey stated that she felt she was a part of such in-groups.

The existence of such groups is especially apparent during professional meetings, but activities persist throughout the intervening months when scientists withdraw to their laboratories. The groups are varied and often change rapidly, but they can be roughly categorized as (1) *clubs* that are often dominated by older men who have made significant contributions to science and who have assembled a support group of former students or associates, or (2) *fraternities* that are usually composed of younger aggressive postdocs and junior faculty, with membership largely dependent on outstanding ideas, enthusiastic discussion of these ideas, and exceptional research performance.

PERCEPTIONS

One common perception about these in-groups in science is that they will be almost entirely dominated by men. Their prevailing attitude is that they may allow a few women participants, if they keep to the periphery and are not too aggressive. Some additional perceptions are these:

- The networking system that is an integral part of the

clubs rarely extends consistently to female participants—females are frequently overlooked or ignored.

- Some women may establish their own clubs with their
  own networking. These often develop when one or a
  few superstar females are identified within a narrow
  specialty area.

- The female scientist may be included in the club as a
  novelty, but not as a prime mover, or she may be
  included because of her willingness to take charge of
  the after-hours or the between-hours social activities
  of the group.

- A female may participate in club meetings if she is a
  protégée of an authority figure (usually a male) or of a
  member of the club's inner circle, but rarely as a
  scientist in her own right.

REALITIES

The social infrastructure of science external to the laboratory and classroom is dominated by clubs, fraternities,
and scientific societies. All provide a system of networking
among professionals. But these groups also serve other
needs, such as a vehicle for informal recognition of success
in research, a vehicle for exposure of frauds, a support
group for scientists, and a platform for informal expression
of opinions about the worth of recently published contributions to research.

Membership in the clubs and fraternities of science is
based principally on demonstrated competence in a specialty area, but with added nonfunctional considerations—

including gender—especially in clubs dominated by male authority figures. Our questionnaire results clearly support the popular perception of a persistent but diminishing gender bias in the clubs of science and, probably as a consequence of this bias, a widespread distaste among women scientists for such groups—even though many admit to being occasional participants in them. Recent trends have been toward reduction of subtle or overt exclusionary practices toward women scientists; this is most apparent in fraternity meetings and less so in the clubs, probably because the fraternities consist of younger professionals who have entered science during a period of feminist–mediated changes.

National and international meetings of scientific societies provide venues for many ad hoc meetings of clubs and fraternities, but they also provide forums for other social encounters. The survey results, and observations made at technical conferences, disclose a noticeable increase in the prevalence of women as session speakers, as members of organizing committees, and as officers of societies. It is also possible to discern increasing female participation in (and occasional domination of) late evening discussions in hotel rooms and bars at those same meetings. So females are no longer only on the fringes of professional groups; they are more and more integral ingredients of them.

A curious account by one of our respondents follows:

NATO Conferences (week-long cloistered sessions for 50–100 specialists) are well-recognized forums for the exchange of the most recent research findings. At a 1991 session on physical chemistry one of the co-organizers was female. Additionally, 25 of the 100 scientists selected as participants were female. Three nights into the session, one of the senior males, a guru in fact, Dr. Michael Zorn, queried a recent postdoc, Dr. Betty Murray, at the bar. "What's with all these women?" "The women are always talking with each other and seem so unapproachable by men." By the end of the week-long session, Dr. Zorn confronted Dr. Murray again, but this time he admitted that he missed the "good old boy" days when any woman present would be flirtatious and adoring in order to be admitted to the cluster of males. "Now women are so confident, independent, and matter-of-fact." Moreover, Dr. Zorn did state that he hoped that even though women were achieving a critical mass, it would not mean isolation of men and women scientists.

Review committees for grants and contracts—a favorite device of some funding agencies—can become a specialized subspecies of the club, especially if the members are reappointed repeatedly by the agency. What emerges is often a persistent core of good old boys and their disciples, who can have surprising power in determining who gets funded and even how lavish that funding will be. Administrators in some of the larger scientific support organizations (such as NSF and Sea Grant) are dimly aware of the problem and may try to provide a periodic infusion of new reviewers, but the inertia resulting from a too-large residue of the old guard is difficult to overcome.

Women scientists participate in review committee activities to a greater extent than previously, but even today it is not unusual for a chairperson to designate, almost automatically, one of them as rapporteur (secretary) for the group. That designation may provide some fleeting visibility, but it also does much to neutralize or at least reduce the contributions that can be made to the discussions by the designee. One positive aspect of this anachronistic situation is that a creative rapporteur is in a position of power over some of the flavor and much of the content of the committee report, regardless of the motivation behind the appointment.

PRESENT STATUS

Some progress has been made in reducing gender barriers to full participation by women in the structured social interactions that permeate the scientific establishment. "Tokenism" is slowly being replaced by more egalitarian attitudes toward the presence of competent female scientists in the informal clubs and fraternities that exist in all specialties and all disciplines. Some indications of shifts in attitudes are these:

- Professional society meetings and symposia provide major extra-laboratory opportunities for social contact among scientists. The more frequent presence of women scientists as society officers and members of boards of directors of societies helps to ensure access to the social interactions that accompany those power positions.

- Sexual overtones, innuendoes, and appraisals still surface during such social interactions, but their frequencies are diminishing and are less often tolerated by female professionals, who insist on recognition as scientists and not as sexual objects.

- "Networking" is a basic ingredient of the communication system of science. Women professionals have long felt excluded from male networks, and the consequences of that exclusion are numerous. Angela Simeone in her recent book *Academic Women: Working toward Equality*[1] lists most of the serious disadvantages, including these:

1. ". . . women are deprived of a sense of community in their work environment and may feel isolated and unsupported."

2. "Women have fewer political allies to lobby for them or their ideas."

3. "Women are less likely to be informed of the latest developments in their fields and to benefit from informal discussion of their ideas and their work."

4. "Women have less influence within their departments and have a harder time being heard by their colleagues."

In other surveys[2] the form of gender discrimination mentioned most frequently was "lack of support from male colleagues at work." This often ranked above other forms of inequitable treatment, such as inadequate salary, delayed promotion, excessive teaching load, lack of time for research, and lack of support from supervisors.

Indeed, inequities still exist, although with some recent signs of penetration. Contacts and interactions with excellent peers and colleagues, regardless of their gender, are clearly becoming operational standards for most scientists.

SUGGESTED ACTIONS

The clubs and fraternities in science constitute conceptually quite different entities. The *clubs* (often correctly termed old boys' clubs) are ruled by a few recognized (and usually older) leaders in a narrow specialty. They are most frequently males; they select the participants; and they preside over the course of discussion. Invitations to participants in their late-evening small-group assemblies are totally informal and deceptively casual. Entry into these salons of the gurus is not simple, but rewards for perseverance can be substantial, once entry has been achieved.

Probably the most direct approach by a woman scientist is to attend as a guest of a guest, preferably one of the inner circle of the in-group. Once inside, of course, future events depend on skill in very shop-oriented conversations held in a cigar smoke-filled room that is too small and whose noise level is too high. Despite the handicaps, this is the environment in which meaningful contributions to the technical discussions are made and in which name and face recognition must be achieved with those present who are important in the specialty area. Brief but relevant discussion with the guru is a requirement to establish an identity and to contribute something sensible if not memorable. The conversation, however brief, should be about the guru's contributions and should never be so long that the guru begins to fidget and to exhibit other overt signs of entrapment. One encounter like this is not enough, though. It must be repeated at subsequent meetings and may be enhanced by interim correspondence referring to the first meeting and asking advice about one's own research, assuming that it is related to the guru's.

Small-group meetings of the *fraternity* are more freewheeling and depend less on gender and more on what individuals are doing in research. It also depends on how effectively it can be discussed in the impossible confines of an overcrowded noisy hotel room or around a group of tables pushed together in a cocktail lounge. Here the leadership is more diffuse, and acceptance depends more on excellence of ideas and the expression of those ideas. A good strategy is to move rapidly through the assembled sub-groups until one is found where the topic is relevant to personal research interests and where some reasonable contribution can be made. The objective (in addition to fast-paced exchange of ideas) is always the same—name, face, and brain recognition. Fraternities tend to be more flexible about applicants than do the clubs, except for the immutable requirements of obvious intellectual ability and apparent competence in the specialty area. Discussions are almost totally in the jargon of that specialty, with a thin overlay of casual conversation that may be partially intelligible to an outsider. The saving feature of fraternity gatherings of this kind is that peer contacts can be made and future one-on-one discussions arranged.

The position of women in relation to the informal groups within science was summarized by Jonathan Cole[3] in the preface to the Morningside edition of his book, *Fair Science: Women in the Scientific Community*, as follows:

> To say that women of science have now entered the central scientific community, and that they have achieved formal equality with men in many measurable ways, does not say . . . that women who have chosen science have an equal chance of ending up in the inner circles of science, nor that they will be equal participants in the "invisible colleges" of the scientific establishment. Resistance to full participation, to the full citizenship of women in the scientific community, continues to exist.

SUMMARY

One aspect of the highly valued collegiality of scientists is the existence of informal groups often referred to as clubs or fraternities—groups that provide the structure for networking. The clubs are often dominated by one or more older men who are considered authority figures; the fraternities are more egalitarian, dominated by those with the quickest wits, the greatest technical perceptions, and the best records of scientific productivity.

Women scientists have been largely excluded from these groups until recently and have therefore suffered professionally because of poor communication with their male colleagues. Many women scientists express distaste for the "old boys' clubs," but recognize their networking value.

Current membership in the clubs and fraternities reflects gradually changing social attitudes among scientists, with greater acceptance of women professionals as colleagues and with virtual disappearance of many exclusionary tactics that have been so frustrating to them. The comfort level between male and female participants is still, however, far from ideal, and some resistance to full scientific citizenship for women persists.

REFERENCES

1. Angela Simeone, *Academic Women: Working toward Equality.* (Bergin and Garvey, South Hadley, MA, 1987), pp. 84–87.
2. Mary L. Spencer and Eva F. Bradford, Status and needs of women scholars, in *Handbook for Women Scholars: Strategies for Success*, eds. Mary L. Spencer, Monika Kehoe, and Karen Speece (Americans Behavioral Research Corporation, San Francisco, 1982), pp. 3–30.
3. Jonathan R. Cole, Preface to the Morningside edition, in *Fair Science: Women in the Scientific Community* (Columbia University Press, New York, 1987), p. xvii.

CHAPTER 11

PERCEPTIONS AND REALITIES

*A catalog of beliefs and perceptions: male scientists' views
of female scientists; female scientists' views of other female
scientists; male administrators' views of women scientists;
female administrators' views of female scientists; female
scientists' views of administrators; graduate students'
views of female scientists; public perceptions of women
scientists.*

Our perceptions influence our attitudes and actions, and
nowhere in human history is this more evident than in how
men view and treat women. Aristotle wrote in the fourth
century BC that "the female condition must be looked upon
as a deformity. . . ." Pericles' ideal woman was ". . . she
who is least talked of among men, whether for good or bad."
What has happened in the intervening centuries? Has some
evolution in thinking taken place?

Differences between *perceptions* and *realities* sometimes
become extremely fuzzy, even in books like this one that

purport to have some factual base. If we accept this possibility, it seems fitting at this point in our narrative to review the whole matter of *perceptions of the role of women in science,* beyond the limited and usually chauvinistic listings given in the introductory sections of some of the chapters. To make some sense of the topic, we have subdivided it into seven principal sections:

1. Male scientists' perceptions of females in science

2. Women scientists' perceptions of other females in science

3. Male administrators' views of women scientists

4. Female administrators' views of women scientists

5. Female scientists' views of administrators

6. Graduate students' perceptions of women scientists

7. Public perceptions of women scientists

With so much of the text behind us, we should have acquired by this time some perspective, some understanding, and maybe even a few insights about the realities of careers in science for women—at least enough to begin sorting out and discarding the mythical elements.

Early in this consideration of perceptions, we have to address some fundamental questions, such as, "Are there biological, intellectual, or behavioral differences between males and females that might account for some of the perceived differences in performance in science?" Simplistic answers would be "No" for intellectual differences and "Yes" for biological and behavioral differences. These answers must be abundantly qualified because of high degrees of individual variability and intrusions of modifying factors that obscure attempts to generalize. It is just such fuzziness that can create and nurture the gender-related perceptions to be explored in this chapter.

In a symposium on "Women and the Sciences," held at Washington College in 1988,[1] Estelle Ramey, a noted women scientist and feminist, listed some biological differences between males and females that may account for their different histories:

- Pregnancy and the desire to nurture young
- Existence of a different sexual biology in males and females—the male being required to achieve erection and ejaculation, whereas the female can be more passive
- The increasing fragility of males with advancing age, as stress hormone levels outdistance testosterone levels

She then went on to make her real point: that even with these *biological* differences, overriding *individual* differences force us to ask, "Which man and which woman?" She further emphasized that, even today, society's expectations for women are different from those for men.

Dr. Ramey also summarized *her* views of male perceptions of females in science. Included were these interesting observations:

- "Attitudes and behavior of women in science are in part consequences of thousands of years of conditioning to the idea that the female is inferior."

- "The only reason females of earlier generations succeeded in science was because they were helped by males."

- "The reason for recent advances in women's status in science is not feminism, but the need for *workers* — especially good scientists."

- "Women are perceived as threats to men only when they occur in numbers; individual or token females are not problems."

This is obviously a very limited listing; we can and will improve on it substantially in the next section of this chapter, based on interviews with male scientists during the course of our study.

MALE SCIENTISTS' PERCEPTIONS OF WOMEN IN SCIENCE

Recognizing extensive variability and even some possible dishonesty in responses, the current male perception seems to be one of positive changes in female roles in science. Many male scientists reported that they now see:

- Less distinction between genders—people are more and more recognized for ability, competence, and productivity, rather than for gender.

- Substantial increases in "comfort levels" when male and female scientists interact.

- Disappearance of doubts about the sincerity of women scientists.

- Dwindling evidence for gender stereotyping of women scientists.

- Obvious increases in the relative numbers of women scientists.

- Lessened need for women scientists to prove merit simply because they are female.

- An evolution in the attitudes of *older* male scientists as well as the younger ones toward greater acceptance of equal roles for males and females in science.

In the relative seclusion of their offices or faculty lounges, however, *some* male scientists will still admit to other perceptions of female scientists:

- They may look upon them as *unfair competitors*, who use their minority status to win positions and their "charm" to get their own way.

- They may retain a pre-1960s negative image of women as "dilettante scientists" and "husband hunters" not fully committed to their profession and not honest in their pursuit of science.

- They may still see women in specialties closely related to their own as distinct threats to their success and well-being, and may use any stratagems—ethical or unethical—to undermine the competitors.

- They may be willing to move in on a promising research area being explored by a woman, expanding on it and dominating it, with scant credit to the female originator.

- They may be reluctant to accept women as *equal part-ners* in research teams, preferring to cling to the stereotype of women as technicians and not as conceptual or quantitative thinkers.

- They may admit *in extremis* that women scientists are still viewed by some men as *sex objects*, to be appraised sexually as well as professionally.

- They may question the abilities of women scientists to lead effective research groups.

An interesting insight elaborated in detail by Vivian Gornick[2] in her book, *Women in Science*, is that male scientists have a fundamental conviction that their actions—even those unrelated to the practices of their trade—are guided by *intellectual objectivity*, which is a cornerstone of their professional existence. Thus they must refuse to admit that any of their actions may be discriminatory or irrational. They have difficulty accepting the reality that they are simultaneously scientists and human beings, and that their behavior just might be modified at times by the same social forces that influence all members of the species.

Some American scientists perceive changes in attitudes of their colleagues toward their wives who are also professionals, citing instances of more supportive attitudes about wives' careers. Examples include an active role by the man in child care and serious considerations of career advantages and disadvantages for the wife in any contemplated move. Some European male scientists report similar examples among their colleagues, but we have to wonder whether discerning trends from too few examples may contain an element of nonscientific generalization—even if made by scientists.

FEMALE SCIENTISTS' PERCEPTIONS OF
OTHER FEMALES IN SCIENCE

The subject of women's perceptions of other women in science needs to be approached very delicately because of the expected high degree of variability in responses. Some questions that were asked during our interviews included:

- "Are the expectations of female scientists about the achievements and productivity of their female colleagues or peers abnormally high?"

- "Is there any element of cooperation or mutual support among women scientists based on similarity of gender?"

- "Do women who succeed in science continually feel threatened, and therefore respond by discriminating against other women?"

- Phrased differently: "Do women scientists feel good and applaud other women who make some major contribution, or is the first reaction one of envy?"

We can offer some very tentative insights about these and related questions, but generalizations are virtually impossible, and even the insights may be labeled controversial or naive.

- Most male scientists are raised in a society that favors the team player. Even in defeat, men can still feel good about a group effort, whereas women feel distraught and alone. Few female scientists to date have been brought up with this team ethic, so it is difficult for them to share victories and defeats with colleagues and peers.

- Women scientists use different criteria in selecting female cooperators or colleagues for research groups. The women look at the potential for developing friendships with the people chosen, as well as at their technical qualifications. Male scientists are less apt to use this criterion. Responses to our questionnaire indicated overwhelmingly that gender was not a significant consideration used by women scientists in selecting associates or assistants, but, from casual observation, the reality seems to be that research groups headed by women often consist mostly or entirely of women.

- Some women scientists find it easier to forgive or minimize the weaknesses of other female professionals, more so than they do for males with similar deficiencies. Other women scientists—maybe more of them—react in the opposite way; they expect female colleagues to be *better* than men, and they can be dismayed when the reality is less than the expectation.

- Women scientists who feel excluded from male networks tend either to isolate themselves from other scientists, male or female, or to form networks consisting of other women with similar feelings of exclusion by predominantly male groups. Networks may be informal and focused on a specialty or they may be formal, such as the numerous women's committees within the structures of scientific societies (see Appendix A). Many women, however, do participate exclusively in the mixed-gender social infrastructure of science, denying the presence of selective forces not related to professional competence.

- Women who are scientific supervisors/managers are often judged severely by other professional females within the group, some of whom may not find the anticipated benefits of gender-relatedness or may feel actually pressured to be unreasonably hyperproductive, simply because of the leader's need to prove herself constantly. Whereas many women in research teams admit that they feel more comfortable about approaching a female supervisor/manager in matters of promotions, salaries, and work hours, many others admit privately that, given a choice, they would rather work for a man, for a variety of reasons, principally because this arrangement is the norm and therefore least stressful.

- Women who are scientific supervisors may also be judged harshly by female technicians or assistants within the research group—again because the supervisor is perceived as failing to provide the expected advantages of gender-relatedness in judging work performance. Women who are support staff members may resent issuance of specific work orders, or insistence on precise hours of work, or maintenance of rigid work habits and schedules, or correction of mistakes if such information comes from a female supervisor rather than from a male. Assertive, experienced supervisors can deal with these tendencies directly and forcefully, but too many inexperienced supervisors find real headaches in coping effectively with the problem.

- Women professionals in research groups may be very sensitive about and resentful toward any inadvertent attempt to lump them in with female technicians or assistants. They may, for example, insist on distinctive name badges or lab clothing and they may respond quickly and negatively to overfamiliarity, such as use (especially by male colleagues) of their first names or failure to address them by their academic titles if they have doctoral degrees. The abandonment of courtesy titles in favor of discordant and premature use of first names is of course a more general societal problem in the United States today and is largely unrelated to gender. Interestingly, the principal culprits are health care professionals—physicians, dentists, therapists, psychologists—who seem to use first names as a device to establish superiority over patients. In general, scientists don't need this kind of game playing.

- A new subset of the activist scientist has emerged— the feminist scientist—who, to varying degrees, has become a partisan in the struggle for equal participation and recognition of women in science. For the more radical ones, this has sometimes meant less time for science in the struggle to meet feminist political goals. For many others, refusal to ignore or tolerate inequities has been a course of action, without reducing their scientific productivity. For still others, sympathy for the objectives of feminism exists, but not active participation. Whatever the extent of individual commitment, new perspectives have been forced on the system of science by the feminist movement. Questions about the status of women professionals are less likely to be ignored or rejected, and an atmosphere of at least lukewarm consideration of the issues raised by feminists prevails.[2]

Another attempt to pigeonhole women academics with reference to women's culture was made by K. Jensen.[3] She perceived three major groups:

1. ". . . women (who) saw a clear distinction between their personal and professional lives . . ." and ". . . claimed no female role models and no affiliations with other women."

2. Women who "assumed a male orientation towards their work, except for the added responsibility of home and family, for which they expected no special treatment from their institutions."

3. ". . . women who called themselves feminists, saw themselves as female role models, belong to women's networks, and work towards an integration of the personal and professional in their work."

Most of the academic women interviewed for Jensen's study (42 individuals) were classified as category 3, despite extensive individual variability. Even though we did not initially attempt to categorize our interviewees in the same way, it seems that a generally similar conclusion could be reached from our survey as well.

MALE ADMINISTRATORS' VIEWS OF WOMEN SCIENTISTS

Any worthwhile basic training course for new administrators during the past two decades has included obligatory seminars with euphemistic titles such as "Managerial responsibilities in equal employment practices," "The managerial role in fair labor practices," "Minority employees and their problems," "Discrimination and its consequences," and even "Sexual harassment in the workplace." With all this sensitization, male science managers are understandably very careful about any expression of opinion or position about women as minorities in science, except for the expected rote statements of relevant institutional policies. Persistent questioning usually leads to "stonewalling," but occasionally small nuggets of real information turn up in the sieve. We have recorded a few, even though our sample size was limited.

- Many male science administrators seem genuinely committed to equal treatment in hiring and promoting women scientists, within the limits of their authority. They quickly point out two problems, though: the conservatism and possible biases of faculty senates (in the case of academic institutions) and the numerically limited pool of qualified women at any level above the introductory one.

- Some male science managers admit difficulties in meeting special demands and requirements of some women scientists, usually in the form of flexible work schedules, unanticipated absences, extensive time off for personal or family reasons, requests for subsidized advanced training, long leaves of absence—an entire litany of departures from expected work activities that seem to be more common among women staff members than among men and that complicate managerial existence.

- Some male science administrators admit a secret dread of "discussions" with feminist activists, who may occupy a good part of the workday with detailed and repeated demands in areas beyond the manager's purview and who are really there to confront and not to discuss. The situation is especially complex if the feminist activist is there as a spokesperson for a women's committee or similar organization, since almost invariably the manager will have limited authority and will be portrayed as the bad guy. Some male managers fear that the discussion may degenerate into accusations and threats of legal action. They explain that such a situation leaves them totally frustrated and hostile, even if they were initially in favor of reaching some compromise.

- Some male science managers—especially the newer ones—have trouble coping with what they describe as the emotionalism of women staff members, especially apparent in times of family stress or when work performance evaluations contain negative components. This may be in part because male staff members have learned to be more effective in hiding emotions during such periods, or in part because women often seek to talk about their problems at length before they are ready to listen to any solutions that might be offered by managers. Inexperienced managers seem to be too quick to offer remedies or suggested courses of action before a problem has been fully explored to the satisfaction of the female staff member. Some male managers are also totally incapable of dealing with tears, and because of this they can leave a mistaken image of coldness or unconcern.

- Male science managers identify special problems with some younger women scientists, especially those whom they feel become overly cynical and hostile too early in their careers. There are certainly enough remnants of inequalities in some institutions to justify concern and action, but most changes do not occur overnight and expectations of "instant gratification" (supposedly a characteristic of the younger generations) can lead to disappointments. The core of the problem seems to be that perceived failure of the institution to meet great expectations can lead to increasing cynicism and accompanying inability or unwillingness to perform at maximum levels—at a career stage when much is expected. Managerial approaches to this kind of problem are difficult, to say the least, beyond a limited amount of counseling.

- Menstruation, pregnancy, and menopause, each representing normal functioning of seasons, cycles, and change in the biology of women, are often touted by males (and some females) as excuses for suboptimal productivity in day-to-day performance. Although this may be less apparent in professional and scientific circles as compared to the general work force, it does indeed exist. Menstruation, pregnancy, and menopause have each been used by men to put down females. Good administrators are sensitized to these gender-specific jibes and try to dampen discussion or "jokes." They realize that the vast majority of women professionals do not permit menstruation, pregnancy, or menopause to diminish their performance.

One of the authors overheard two high-ranking male administrators, both directors of laboratories, engaged in conversation about menopausal women.

In supposed "private" conversation, both directors admitted that they felt they were able to discern when women were experiencing the "change of life" and had learned to live through the associated irritations. Both then confessed that their "favorite" employee subpopulation was postmenopausal women. They claimed that this cohort satisfied all the new EEO game rules, provided guidance, seniority, and stability for the junior females, and performed under less than ideal conditions with gusto. They went on to comment what a "relief" it is to have women who are past all of the controversial demands such as maternity leave, babies in the laboratory, child care and educational programs. Both concluded that they would take five postmenopausal women to every one premenopausal woman! Whether these same administrators would prefer male scientists whose sexual lives had peaked many years earlier did not come up in their conversation.

FEMALE ADMINISTRATORS' VIEWS OF WOMEN SCIENTISTS

Most female science administrators reach their positions through a combination of competence, energy, and luck (as is true of most male science administrators). They inhabit a rough, competitive world in which few mistakes are tolerated and are under constant pressure to demonstrate their competence. They have the added burden of being invaders in what is still considered a man's turf, and occasionally feel the need for allies among the women scientists in the organization. Female bonding is a suitable description for this kind of alliance, except that it may be complicated by a boss–employee relationship as well.

A principal problem is that female staff members may assume that a woman administrator will automatically (and vigorously) champion their side, regardless of what the best course of action may be. Any failure to perform in accord with that assumption can be construed as a "sellout" to the male establishment. A difficulty here is that not all female administrators are active feminists, but they may feel forced into that role by expectations of the staff. The job may become a balancing act, trying to keep the best interests of women in mind, but not allowing too many gender considerations to enter the decision-making process. Some modicum of hostility toward women staff members who expect partiality can be an unwanted consequence.

A woman administrator can feel very lonely at times: when she enters an important management meeting and finds that she is the only female present, or when she must vote unfavorably on a promotion list that includes several women professionals who are also friends, or when she attends a week-long managers' retreat and feels somehow excluded from the easy give-and-take of male cohorts. At these times, the support from and the bonding with women professionals in her group can seem very remote and ineffectual.

There is, however, a positive side to the relationship of a woman administrator with female professionals in the organization. Friendships with compatible, understanding individuals and even forms of mentor relationships with senior staff members can make the job tolerable and enjoyable. Good advice and suggestions, freely given, can sometimes prevent serious blunders or broaden insights on a problem. One-on-one lunches with a perceptive colleague can serve as excellent safety valves during particularly strenuous days. Of great importance, too, is daily association with intelligent, informed women professionals—interesting people who are delighted with their scientific careers and who exhibit a high degree of self-motivation.

FEMALE SCIENTISTS' VIEWS OF
MALE ADMINISTRATORS

Female scientists must remain on guard and be active in the selection process for male administrators. It has been the experience of many women scientists included in our interviews that they cannot assume that a male administrator is fair in all decisions if he boasts of having had a female mentor or has made a clear pro-female choice at some time. This is important because even when clear and just equal opportunity policies are in place in an organization, the boss has subtle ways to put down females if he so desires. The experience of Dr. Thelma Watt serves as an example:

> I have worked in the same nonprofit research foundation for the past twelve years, supporting my research with federal and state grant and contract awards which gives me maximum independence and flexibility. Over this time I have had two directors. The first one had a clear pro-female attitude, resulting in nearly a 50:50 split between males and females in our foundation. EEO was never an issue, and the women spent little time even discussing inequities. We considered the workplace environment to be positive, comfortable, and enlightened. Every woman scientist was consulted about laboratory directions, new hires, opinions and ideas for change. Her ideas were honored and respected. The policy was firmly in place and our laboratory served as a model for other institutions. Unfortunately, that director retired. We were naive to assume that things would remain the same.
>
> The replacement director was always pleasant and kind to women scientists at national and international meetings, thus we assumed that he would follow in the footsteps of the past director. Had we looked carefully into his record of performance, we would have gleaned a sexist past including that it was important to him that his wife stay home and care for their children; this alone sends the message that the woman's place is in the home! When a woman scientist's family support system

temporarily collapsed, the new director had no tolerance. If a parent needed to leave the laboratory to get an injured child from school, the director was annoyed. In a few short weeks the lab went from one of equity to one fueled by testosterone. Within two years, several of the women were fired, resigned, or moved elsewhere. In short, strong institutional policy plus a sympathetic policymaker must *both* be in place to have an enlightened workplace.

GRADUATE STUDENTS' PERCEPTIONS OF WOMEN SCIENTISTS

The microcosm populated by science graduate students within a university intersects constantly with that of the faculty—courses given, seminars attended, advice offered on thesis topics, approaches to research discussed, preliminary results of thesis research evaluated, brown-bag discussions held, faculty–student picnics and other social gatherings attended—on and on. With all these interstices, it is not surprising that perceptive graduate students form very definite opinions about faculty members (though these opinions may have inadequate data bases, may reflect lack of experience with faculties of other institutions, and may disclose an extraordinary degree of naiveté about science and scientists in general).

Our survey did include a sampling of female and male graduate students, so some discussion of their perceptions is not unreasonable. The most useful aspects of the interviews, from our point of view, were unsolicited comments about women in science and women faculty members at their institutions. Generalizations that emerged are covered in the following paragraphs.

Today's graduate students—most of them anyway—

are fully aware that they must make all the right moves if they are going to succeed in the present highly selective system of science. Some of the most critical of those moves are choice of thesis advisor, gaining visibility as a leader and producer, acquiring competence in laboratory techniques, extensive immersion in current and earlier literature of the discipline, learning skills of oral presentation, assembling an outstanding academic record, and effective performance as a teaching assistant. The most perceptive students don't just stumble through graduate school; they have a well thought-out game plan from the day they arrive in a university city or town.

Some students enroll because the school has strength in a discipline that interests them; others have been attracted by the presence on campus of one or more senior faculty members with outstanding credentials in that discipline; still others have come because of availability of financial support. Whatever the reasons, graduate students are faced constantly with choices that can shape their future careers, and they know it. One important early choice is that of a faculty advisor—someone with a blend of professional and personal attributes that promises a satisfactory working relationship. Factors to be considered include (but are certainly not dominated by) a gender relationship, especially if women scientists are among the choices.

The majority of responses to our questionnaire from female faculty members indicated that *their* choices are free from gender considerations, and we believe that, but what of the choices made by *students* in this mutual game of "pick your partner"? Most male students don't seem to have a problem with gender in this situation; their preferences seem to be based on other factors. However, some male students deliberately avoid selecting female advisors.

Female students—some of them at least—gravitate toward female advisors, if other factors are equal. This may result from a sense of greater security and easier communication with a woman advisor, a felt need for a female role model, a search for a mother substitute, or none of these.

Some additional comments need to be made here, though. Many graduate students are thorough pragmatists, conditioned to examining all the angles before making any move. They see the disparity in numbers of women scientists in senior positions and hear about continuing inequities in acceptance of females into male networks, so a small amount of unease can be generated over the choice of an advisor. "Will a woman advisor be able to give me the contacts and open the necessary doors as compared with a man?" "Will I as a graduate student be subject to negative effects of any lingering and well-concealed biases of male committee members during qualifying exams and thesis defense?" "Will the recommendation of a woman scientist have the same weight as that of a male 'club' member when the time comes to find a first job?" These considerations are well down on the priority list for most graduate students— far below professional considerations—but they do exist and may result in a decision to avoid a woman faculty member as a mentor/advisor.

Some graduate students are definitely macho types, sometimes with vaguely sensed and societally induced feelings of male superiority. The new environment, with much closer and more informal faculty contacts, combined with a greater freedom of action and expression, can create problems for women faculty members. Lack of proper respect and overfamiliarity are the more obvious signs, as is continued and unnecessary probing of the limits of the faculty member's knowledge, all indicators of persistent immaturity. The problem can be more severe for younger female faculty members, who are not separated from graduate students by a large age barrier; here the concept of "professional distance" needs to be enforced. Interestingly, some otherwise mature male students may not even be aware of their behavior until they are taken by the ear and talked to directly and forcefully.

Other than some preference by female graduate students for female advisors, gender considerations enter professional academic relationships between student and teacher only peripherally, except for the important role model considered earlier in this book. Female graduate students *expect* women faculty members to be exceptional, and they can be very critical, if those expectations are not met—possibly even more critical than they are of less than perfect males. Failure to perform outstandingly can somehow be considered a reflection on all professional women, rather than an expression of individual capabilities.

PUBLIC PERCEPTIONS OF WOMEN SCIENTISTS

We're extending well beyond the boundaries of our survey in this section on public perceptions of women in science, but some treatment of the topic seems relevant. Scientific data are used as background for societal and political decisionmaking, and more and more women are contributing to the information base used in making these decisions. How does the average lay person view women scientists, and what factors are influential in shaping those views?

The average newspaper-reading, television-watching nonscientist is at best only dimly aware of the great disparity in numbers of female versus male scientists, especially in the more quantitative disciplines, but he or she has been accustomed to reading about and seeing male authority figures. Male scientists are described in feature articles about new discoveries; male scientists are quoted making pronouncements about technical matters of public concern; male scientists are interviewed when ecological disasters occur; and male scientists dominate public television outreach programs. Should there be any reason, therefore, not to expect a small element of surprise when a woman scientist is quoted, or photographed, or interviewed as an authority figure in her specialty area or as the investigator responsible for a significant discovery? Surprise may be accompanied by other reactions: questions about whether the woman scientist is merely acting as a spokesperson for a group consisting mostly of men; questions about how she was able to reach this level of visibility and credibility in a traditionally male-dominated occupation; and questions about the changes in society's attitudes that must be embodied in her public performance. Part of the reaction must also be a secret pleasure, mixed with a touch of envy, on the part of other female professionals as well as secretaries and homemakers, when a woman "makes it" intellectually and financially.

A recent phenomenon in American television has been the popularity of daytime talk shows hosted by women, as a partial respite from a stultifying combination of interminable "soaps" and puerile game shows. These talk shows, undoubtedly because of the insistence of the female commentators, provide greater visibility for articulate, knowledgeable women scientists, speaking in their specialty areas. It is true that the principal audience is also female, but the public exposure must be salutary in supplying, at least to that segment of the public, a needed perspective on the important role of women in science. Not many men watch these programs, which is unfortunate, but more and more women scientists are gaining critical media experience and visibility. Additionally, many more women are now reporters of stature, so men as well as women see them on the nightly news.

Another recent phenomenon, apparent in all the news media, is the increasing frequency of women scientists who are asked to voice their opinions on environmental matters. What we seem to be seeing is a greater number of well-informed women professionals who are actively participating in research and teaching in environmental specialties. They are willing to take public advocacy positions (possibly even more willing than most male counterparts). One female interviewee attributes this upsurge to the fact that women have less to lose. On the other hand, the reverse could be said if their opinion is counter to that of the majority of scientists. This is a rough arena, populated by such climax predators as lawyers, politicians, and industry consultants. Successful participation requires expertise, experience, and credibility. The combination of scientific competence and social skill may be contributing to this phenomenon. At any rate, women professionals are present in significant numbers, and the public is paying attention to them.

As long as women scientists flock to all areas of science in addition to more public ones, this kind of exposure of women as authority figures can have only a beneficial effect on women in science. The subliminal effect of women treated as authority figures of the first rank through the mass media cannot be overstated. The only problem is when a public-type position short-circuits the career of a brilliant researcher. Nevertheless, the two, like a profession and motherhood, can be balanced.

Women scientists have been likened by some males to an introduced exotic species, not native to the habitats of science, but with highly adaptive capabilities. The men indulging in this mildly chauvinistic comparison have elaborated further on the concept: just as with deliberately introduced organisms of any kind, a "code of practice" had to be developed to ensure that risks of deleterious environmental effects and overpopulation were minimized. Such a code, as developed for animals and plants, included a sequence of actions, beginning with examination of the species in its native surroundings, determination of its ability to interbreed with native species, and severe limitation on numbers of individuals allowed entry. These same men see many parallels with the unwritten rules governing admission of women into science—beginning with restricting participants to small, early controlled introductions and moving on to assessing survival potential in a harsh new environment, studying competitive behavioral characteristics that might be harmful to the natives, and observing population growth potential under stressful conditions.

The code of practice has been adhered to and the introduction must be considered a qualified success: women have survived the imposed stresses and now occupy a slowly expanding ecological niche in the jungle we call modern science.

SUMMARY AND SUGGESTED ACTIONS

There are many perceptions about women in science which are prevalent, but which may or may not reflect *reality*. What is desired is a working environment that is free from constraints for all persons, regardless of gender. So it would seem that at this point some recommendations for the future reality are in order.

The following recommendations were repeatedly brought to our attention by various persons interviewed.

- Schedule professional business, growth, and evaluation during normal working hours (not evenings and weekends), and hold these in the work environment.

- Examine draft media releases and promotional materials and ensure equal treatment for men and women; incorporate gender-free language.

- Make sure that equitable treatment of women and minorities becomes part of the formal reward criteria for promotion at your institution.

- Immediately take steps to correct inequities which may emerge during evaluation of candidates or faculty. There should be no differences in the way the background and qualifications of individuals are discussed or acted upon.

- Conduct after-the-fact interviews with short-list candidates who are not hired to detect whether they feel they have been treated differently than others.

- Keep a personal journal of instances of differential treatment that you have witnessed.

- Maintain a positive philosophy.

- Focus on one or a few target areas of dissatisfaction, yet set a limit on how proactive you will become. Being a good professional scientist must always be first priority. There is no way to fight all of the battles.

- Whenever possible, use humor.

The Association of American Colleges (AAC), through its Project on the Status and Education of Women, has also focused on a number of issues that confront women in science. In its 1986 publication, *The Campus Climate Revisited: Chilly for Women Faculty, Administrators, and Graduate Students*, Bernice Sandler makes several dozen recommendations about creating an environment that will stimulate women's achievement and success.[4]

REFERENCES

1. Estelle R. Ramey, Keynote Address, Minisymposium on *Women and the Sciences: Expectations, Realities, Hope* (University of Maryland, College Park, MD, 1988) (unpublished).
2. Vivian Gornick, *Women in Science: Portraits from a World in Transition*, (Simon and Schuster, New York, 1983), pp. 34–38.
3. K. Jensen, Women's work and academic culture: Adaptations and confrontations, *Higher Education* 11, 67 (1989).
4. Bernice Sandler, *The Campus Climate Revisited: Chilly for Women, Faculty, Administrators, and Graduate Students*, Association of American Colleges Report (Washington, D.C. 20009).

CHAPTER 12

SUBTLE FORMS OF GENDER-BASED DIFFERENTIAL TREATMENT IN SCIENCE

*Subtle gender-based differential treatment, from
backlashing to tokenism, with some attention to body
language, condescension, devaluation, exclusion, hostility,
invisibility, role stereotyping, and sexual innuendo.*

We doubt that any sensible person would deny that women have been subjected to overt discrimination throughout the history of science. Until recently, the scientific community was populated at professional levels almost exclusively by men. Women were traditionally in the background, in supporting nonprofessional roles. Substantial signs of change have been seen in recent decades; overt discriminatory practices that largely excluded women from the professional ranks have diminished or in some instances disappeared. As we've noted, governmental and institutional actions, combined with the momentum of the most recent feminist movement, which began in the late 1960s, have contributed to the change.

But, a point often overlooked and usually denied is that whereas *overt* discrimination has declined, a large residue of *subtle* discrimination—individual and institutional—still persists. These less tangible attitudes and practices reflect an imperfect adaptation of the scientific community to the concept of gender equality. They also present challenges more difficult to confront and eradicate by legal and other means because they are ephemeral and defy the usual confrontational approaches.

We will try, in this chapter, to categorize some of these persistent subtle forms of discrimination against women scientists, as described during our interviews and discussions, and as reported in earlier studies, notably those of Eileen Shapiro.[1] Even though some may seem minor, and in a few instances may seem tinged with just a soupçon of paranoia, they do form part of the totality of existence for at least a segment of the women in science. The categories described by Shapiro as "nonactionable discrimination" encompass most of the broad range of activities that constitute subtle discrimination. We have adopted them, and added a few of our own, to prepare the following list:

- exclusion
- condescension
- role stereotyping
- tokenism
- hostility
- sexual innuendo
- invisibility

- body language
- backlashing
- devaluation

We are convinced that these subtle discriminatory practices and attitudes are still aspects of everyday life for women in science, and that they can contribute significantly to the lack of joy or acceptance that some women feel in their careers.

EXCLUSION

The bases for decisions affecting the careers of women scientists have been explored in earlier studies. For example, Jessie Bernard[2] described criteria for such decisions as either functional (intelligence, training, and experience) or nonfunctional (race, religion, extent of commitment to careers, and other devices to disguise bias). The point is made that a woman's credentials have been often judged by nonfunctional or functionally irrelevant criteria, even in instances, such as hiring decisions, where functional criteria dictated clear choices. Exclusionary practices of this kind are rarely overt, but they may be subjective components of the decision-making process, when the decisions are being made by men.

We have already considered, in Chapter 10, what is probably the most professionally damaging form of exclusion of women scientists: the lack of participation in male networks, especially those concerned with exchange of ideas, support for new research proposals, acceptance of graduate students, tips on job openings, and collaboration in ongoing research. The barriers seem to be less severe now than formerly, but *access* to all the informal but not unimportant give-and-take of scientific networking still seems to elude many women professionals. Detrimental effects on careers of this continuing lack of access are often intangible and usually impossible to quantify, but few would deny the value of participating.

An interesting phenomenon illustrating a form of exclusion from male networks was described by Jonathan Cole[3] in the preface to the Morningside edition of his book, *Fair Science: Women in the Scientific Community.* He points to the frequency with which an altered pattern of discussion occurs when male colleagues converse with women scientists. They "talk shop" with other men, but switch to nontechnical topics like families and vacations when addressing women. Cole characterized such behavior patterns as issues of "scientific citizenship" that still influence in a negative way the workday experiences of women in science.

Exclusion can take many other forms; one exclusionary practice that is potentially dangerous to the health of women received media coverage in 1990.[4] Amazingly, women have not been included as subjects in a number of major medical studies funded by NIH, despite the fact that findings of those studies have been used as a basis for treatment of diseases that affect both sexes, such as heart disease and cancer. The weak excuse offered was that women's hormonal fluctuations made it difficult to interpret the data! As a consequence of the unfavorable publicity, NIH has modified this bizarre exclusionary practice and has established a new office of Research on Women's Health.

CONDESCENSION

One of the more exasperating (to women) forms of subtle discrimination is based on the anachronistic ideas that women must be protected from harsh realities and that men should handle the more challenging work that requires a high level of competence. These reminders of attitudes that have persisted from the overtly discriminatory early history of science surface occasionally—even today—to the profound annoyance of competent women scientists.

Condescension surfaces unexpectedly and in many forms. It can be as simple as not selecting women for high-endurance field studies because they are too "fragile," or harboring unstated reservations about appointing a woman to an arduous committee assignment involving extensive travel. Some men are vaguely uneasy about the abilities of women to perform rigorous and exhausting series of experiments. Some men may be unwilling to allow women scientists to take equal responsibility for transporting, setting up, and discussing exhibit material for technical conferences. Some men worry about the capability of women as managers of long-term multidisciplinary research projects.

The most distressing aspect of condescension is that it is often unintentional; that many of the men who are guilty of such practices usually have not the faintest perception that they are doing anything that could be considered discriminatory. An insidious and potentially harmful component is that women exposed to enough of these incidents may get very angry. Should they show annoyance in questioning these practices, they may be accused of emotionalism. Worse still, after being patronized for a long time, they may begin to doubt their own competence.

ROLE STEREOTYPING

Regardless of demonstrated competence and professionalism, women scientists can never seem to escape completely from role stereotyping imposed by some male colleagues. Women are expected to be supportive of men, feminine in appearance, often emotional, sexually available, and committed primarily to home and family. Departures from the stereotype can confuse and dismay those male counterparts who are insensitive to the reality that women can be far more complex than men and much less amenable to being placed into outmoded pigeonholes. Role stereotyping places many professional women in a "damned if you do and damned if you don't" situation, effectively described by Angela Simeone.[5] Fitting the stereotype of wife and mother leads to perceptions of lack of commitment to science, and failing to fit the stereotype can be perceived as unseemly aggressiveness and competitiveness.

Some forms of gender stereotyping do seem to be receding, however. One that has been almost universally overthrown is the automatic assumption (by men) that women in a group situation will perform secretarial and service functions—taking notes and making coffee, to name the most common ones. More and more men seem to be becoming sensitized to the fact that this kind of assumption is no longer acceptable to most professional women and may now lead to decidedly hostile responses from women unwilling to act in a subservient manner.

TOKENISM

Encompassing aspects of role stereotyping, but extending well beyond its perimeters, is "tokenism." In tokenism, men seriously outnumber women in positions of stature and impose all their expected female stereotypes on the lone or scarce female representative(s). The tokens must un-

willingly assume responsibility for all members of the female gender, providing exemplary performances in the
many stereotypical roles, as was well described by Angela
Simeone.[5] Tokens are highly visible yet still isolated in an
often suspicious and uncomprehending male enclave.

The token woman is expected, almost automatically, to
assume the twin burdens of continuous visibility and service
as a role model, whether she perceives those jobs as legitimate for her or not. She must endure judgments by male
colleagues about her every action—from manner of dress to
attendance at staff meetings to research productivity—as
somehow typifying the actions of all women. In addition,
she is expected, unfairly, to be the focus for any and all
complaints that men may have about women in science,
whether they are directly relevant to her or not. Sometimes,
too, she may be an unknowing object of the fantasies of
certain male colleagues. The opportunities for high-level
stress in such situations are definitely awesome.

HOSTILITY

Most of the subtle discriminatory practices included
here can be interpreted by some observers as evidence of the
hostility of males toward female invaders of "their" turf. The
hostility can have its roots in fear of being displaced or being
bested by a woman competitor. It can also result from basic
lack of understanding (or acceptance) of role changes for
women, away from the traditional ones of wife and mother,
or even the faceless enduring support figure. Sandra Harding, in her book, *The Science Question in Feminism*,[6] has a
hard-line explanation, that ". . . masculine gender identity
is so fragile that it cannot afford to have women as equals to

men in science." We may question some of the premises for such a conclusion, but we don't deny the subtle indications of persistent underlying male hostility.

Deep-seated hostility is often disguised as "teasing"— the practice of making belittling, inappropriate remarks thinly veiled as humorous. The objects of these attacks—the women scientists—find little humor in this invective, but the practice is often overlooked by other male colleagues, even though they may comprehend dimly the kind of vicious game that is being played.

Unfortunately, other expressions of male hostility are more direct and overt, even though they are diminishing in frequency in today's climate of nonacceptance by women of any level of injury. Some male professionals still espouse the Aristotelian tenet that being a female is an infirmity, and some of those men are affronted by the concept of gender equality, in science or anywhere. They react with anger at what seems to them as the humiliation of seeing women professionals not only performing as well as men, but in some instances outperforming men and even dominating scientific forums. Quiet rage may erupt as public displays of hostility, if circumstances provide the opportunity. Other more blatant expressions of incompletely repressed hostility that women have come to expect from some men include the nasty aside and the inevitable double entendre.

SEXUAL INNUENDO

Undoubtedly every women scientist has her own collection of episodes and incidents in which she has been treated as a sex object rather than as a professional. The stories that we have heard seem endless in variety: the slide of a woman in a bikini used by a man to conclude an oral presentation at a scientific meeting, the suggestive jokes at late-evening small-group cocktail parties, the unnecessary and unwanted references to appearance or dress offered as part of casual conversations in the workplace. These are just a smattering of the indignities women suffer through when male scientists focus on women as sexual objects rather than professionals.

Responses of women professionals to attempts by men to reduce any conversation to physical/sexual levels are highly varied. Many women act as if they had not heard or had not understood the offensive remark, and thus avoid the issue in that way. Other women can be more confrontational, responding more vigorously with an impromptu verbal counterattack of their own. Some women seize the opportunity to conduct (then or later) a small private session to inform the miscreant about acceptable conduct of professionals in mixed social situations. A few women scientists have been pushed so far that they are forced to resort to well-placed kicks to vulnerable parts of the male anatomy. And a growing number of women are now going to the courts.

The truly astounding aspect of this deviant behavior by males who seem otherwise rational is their *persistence* in their lewd behavior and comments. This is particularly annoying when the male is a contemporary. The situation created is almost certain to be "no-win" for the man, and we're talking here about males with mostly upper-range IQs. How can they be so stupid for so long—sometimes for their entire careers? Good answers elude us, except maybe to remember that intelligence and maturity may not be connected. We recall the quote of Mark Twain from *Pudd'nhead Wilson's* Calendar: "Habit is habit and not to be flung out of the window by any man but coaxed down-stairs a step at a time."

At times it seems almost incomprehensible that otherwise intelligent well-educated males can be so lacking in perception and so coarse in their treatment of female colleagues, yet the reality intrudes with disgusting frequency in any mixed group. The only excuse—and it is a lame one—is that only a minority of male scientists are guilty of the actual commission of these offenses and even they may not always be fully aware of the revulsion engendered by their verbal assaults.

INVISIBILITY

Classic campus vignettes from the past describe the older science professor who never called on female students or whose seating plan actually segregated them in a rear corner of the lecture hall, thereby assuring their almost complete disappearance. But that was an earlier unreconstructed time, when science was principally a man's professional occupation, and women were seen as intruders, providing, unobtrusively, only necessary technical support and little else. Today, women are more visible as center-stage participants, although habits of the past die hard.

Complaints about the persistence of female "invisibility" as a part of doing science were made by a significant proportion of the women respondents to our questionnaire. Examples cited included such minor annoying practices as having a session chair address a mixed audience as "Gentlemen"—or, worse still, adding as a lame and delayed afterthought "and ladies"; having the work in a jointly authored paper discussed as though it had been done entirely by the (male) senior author; having the (female) laboratory staff, during a site visit by a review team, "introduced" by the (male) principal investigator with an impersonal wave of the hand; or having a working group chair persistently recognize male participants and ignore females who want to enter the discussion.

These responses, and many similar ones, are, as was pointed out previously by Eileen Shapiro,[1] extremely prevalent, seldom deliberately malicious, and usually unconscious, to the extent that the perpetrator can be amazed and chagrined when they are called to his attention. They do carry the unspoken connotation, though, that women's contributions can be ignored as presumably inconsequential, and as such the actions are clearly discriminatory.

> In 1975, Dr. Darlene Erik and Dr. Bridgette Brian applied to the National Science Foundation for funds to support their research ideas about the effects of photoperiod and prolonged darkness on plant ecophysiology. The research proposal received good marks and was funded. The two women were gratified. Yet, when the peer review verbatim comments were returned to the authors, they were disconcerted to read such comments as these: "Having these scientists with their lower salaries conduct this project should mean that NSF will get a big bang for its bucks." "I expect these researchers will be successful in this interesting endeavor . . . and we know they will work twice as diligently as men."

BODY LANGUAGE

Most of us have at some time thumbed through a paperback pop-psychology book that attempted to interpret physical actions as reflections of unexpressed attitudes or feelings. Women scientists are usually familiar with all kinds of variations of these nonverbal responses of male colleagues to their participation or even to their presence in a group. Some male faculty members are remarkably ill at ease in one-on-one conferences with female students or female colleagues; they avoid eye contact and they never, never shift their gaze below the neckline (even if others fixate there); they clear their throats repeatedly; they fidget with desk toys or writing tools; they pace a lot; and they seem so relieved when the conference is over.

Other kinds of behavior toward female colleagues can also be diagnostic: male scientists are careful to maintain adequate distance from females during group discussions, and are extremely diligent about avoiding any physical contact that might be misconstrued as a sexual advance— even such a simple gesture as a handshake. Behavior at scientific meetings can be revealing, too, especially the reactions of some male scientists to paper presentations by female colleagues. Some men may lean back in positions of disinterested repose; some may close their eyes for a short nap; some may study their programs or other handout material intently throughout the presentation; some may use the occasion to dash off a letter or memo; and a few may even lurch to their feet and abandon the session room in favor of corridor discussions of any kind.

It can be difficult for perceptive women scientists to ignore such overt signals of discomfort with their presence or disinterest in their professional contributions. One feeble consolation may be that these same insensitive men will sometimes treat male colleagues with similar or additional rejecting behavioral signs. Of course, not all male body language is negative; many signals can be supportive or even laudatory (although these are usually reserved for other men), but they must be clearly and carefully differentiated from those indicating personal (sexual) interest.

BACKLASHING

The rise of feminism has produced, as might be expected, an appreciable amount of reaction—of "backlashing"—to the movement, to the participant, and even to women in general. Radical feminists are not numerous in science, probably because professional demands are so all-consuming that little time or energy is left to fight other battles. Many of the women scientists included in our study described themselves as interested in the feminist cause, but most often from a passive rather than an active position. Some of the male scientists interviewed in our study expressed sympathy for the goals of moderate feminism, but a few described some of the extremist views and identified possible negative impacts on male careers and on the aspirations of women in science.

The more outspoken of the men pointed to the unfair advantage in hiring accorded to women by institutions with aggressive equal employment opportunity policies, which in some instances dictated hiring women over equally qualified males. Others pointed to the seemingly inordinate amount of their time spent in committees formed to develop fair employment practices and policies. Others were annoyed with the endless demands of women's committees that occupied substantial parts of the agenda of faculty meetings. Still others wondered about the time made available to feminist activists for nonteaching and nonresearch involvements.

These and other concerns have produced a modest amount of backlashing—of displeasure on the part of some men (and a few women) over the whole issue of feminism and the women committed to its success. That displeasure has at times translated itself into deliberate inertia on the part of university administrators and faculty committees in approving actions favorable to women. This, of course, leads to so-called counterbacklashing by women, with further erosion of time available for other professional activities such as teaching and research.

DEVALUATION

One ubiquitous and persistent form of subtle discrimination is *devaluation* by men of the abilities and accomplishments of women scientists. The downgrading may take many forms, such as: attributing a woman's success to chance, or to affirmative action, rather than to competence and productivity; evaluating professional credentials differentially, favoring recruitment of males rather than females; unfairly trivializing or ignoring a woman's contributions to research, to discussions, or to concept development, even when they are equal to or better than those of male colleagues; or subjecting a woman's credentials, publications, or statements to greater scrutiny than those of men.

These signs of the reduced value placed by men on women's achievements in science do show some indications of abatement, but they have by no means disappeared. Successful women may still be described as exceptions to the rule (that women are less competent than men); women administrators may still be suspected of "riding the affirmative action wave"; and women scientists who make significant breakthrough contributions to their specialty may be resented rather than congratulated.

The process of devaluation, like so many other forms of subtle discrimination, can be either merely exasperating or actually very destructive to the careers of women scientists. An environment in which substantial contributions by women may be minimized or dismissed as insignificant or ignored is not one that is conducive to comfort or productivity. As is the case with most other forms of subtle gender discrimination described in this chapter, what is required is a system-wide modification in attitudes of male scientists and administrators, principally in accepting the reality that women professionals are as competent as men, and that their contributions must not be evaluated on the basis of gender.

*   *   *

All of these subtle forms of gender discrimination can provide an added and totally unnecessary burden of stress for women scientists, many of whom are already trying to cope with severe career-related and family-related demands. They have no place in the professional environment; their persistence can only be attributed to a lack of sensitivity and perception on the part of male scientists and an unwillingness to create confrontational situations on the part of female scientists.

SUGGESTED ACTIONS

A legislative foundation now exists to combat overt forms of gender discrimination and it is being used in the courts, but the many forms of subtle discrimination are elusive and are described by lawyers as "nonactionable." This implies that alternative approaches have to be employed to reduce the level of abuse. Eileen Shapiro[1] has listed some of the issues involved in coping with subtle forms of discrimination. She suggests that a woman should first recognize and acknowledge that discrimination has occurred, then try to determine if and how she may have participated in the discriminatory action, then assess the damage done to her as a professional and as a person, and then finally develop strategies for coping with future incidents.

Though the forms of subtle discrimination are varied, it is still possible to propose general strategies. Some responses include these:

• A discriminatory incident can be used in professional and social situations to confront and to educate the perpetrator. Subtle discriminatory acts should be identified as such and rebutted when they occur, preferably in a direct and civil manner and certainly without excessive display of emotions.

• Consciousness-raising, through seminars, mixed-gender workshops, and one-on-one discussions, can be effective in identifying and combating many kinds of subtle discrimination. In most instances, incidents result from unconscious or insensitive behavior on the part of the male, and once identified they will usually not be repeated (at least not in the same form).

• Expressions of underlying male hostility toward women are difficult to cope with. Superficial symptoms such as baiting or belittling can call for direct confrontational responses, but the underlying disease persists and the symptoms will reappear, although possibly in reduced form.

• Women must not allow themselves to be forced into the role of participant in any discriminatory interaction—as, for example, accepting as humor jokes of a personal or sexual nature. Clear rejecting behavior, verbal and nonverbal, is always the response of choice for a victim.

• The time, place, and frequency of responses to forms of subtle discrimination must be chosen carefully by women scientists. Factors to be considered include extent of diversion of energy from other activities, degree of distraction from professional responsibilities, and effects on personal and professional attitudes.

Most of these suggested remedial actions can be considered as *effective communication* in which the female victim tries forcefully to make the male perpetrator aware of his transgressions and offers some instruction in proper procedures. Unfortunately, the most professionally damaging form of subtle discrimination—exclusion—is not amenable to simplistic approaches and requires a fundamental change in male attitudes before its effect can be mitigated. Progress has been slow, but visible.

SUMMARY

This chapter has focused on the many varieties of subtle discrimination that, unfortunately, still annoy, exasperate, and enervate women scientists, often resulting in lower productivity and disenchantment with the social component of scientific activities. The forms taken—exclusion, condescension, role stereotyping, tokenism, sexual innuendo, invisibility, body language, devaluation, and backlashing—can be viewed as examples of male hostility (or at least unease and fear) directed toward women scientists who are invading what has been traditionally male territory. Many of the incidents result from insensitivity on the part of male scientists. Fortunately, most of these men are marginally trainable, so an intensive consciousness-raising session can often produce favorable behavioral modification. Assertiveness training is equally effective for the female scientist.

Emphasis here on subtle forms of gender discrimination in science should not obscure the reality that remnants of *overt* discriminatory practices persist, despite a gradually improving legal screen against them. Inequities still exist in salaries, in promotions to senior academic ranks, and in access to the operational networks of science. Legal measures can address the more obvious abuses, but changes in attitudes of men toward women as scientific colleagues are still required before true gender parity exists. Only then will all the exclusionary practices that have characterized the past finally disappear.

REFERENCES

1. Eileen Shapiro, A survival guide, in *Handbook for Women Scholars*, eds. Mary L. Spencer, Monika Kehoe, and Karen Speece (Americans Behavioral Research Corporation, San Francisco, 1982), pp. 121–122.
2. Jessie Bernard (ed.), *Academic Women* (University of Pennsylvania Press, University Park, PA, 1964), p. 75–77.
3. Jonathan R. Cole, Preface to the Morningside edition, in *Fair Science: Women in the Scientific Community* (Columbia University Press, New York, 1987), pp. xiii–xx.
4. Diana Morgan, Unlocking research barriers, *AARP Bulletin* 31(10), 1 (1990).
5. Angela Simeone, *Academic Women: Working Toward Equality* (Bergin and Garvey, South Hadley, MA, 1987), pp. 78–83.
6. Sandra Harding, *The Science Question in Feminism* (Cornell University Press, Ithaca, NY, 1987), p. 64.

CHAPTER 13

ERAS IN THE CAREERS OF WOMEN SCIENTISTS

Phases in scientific careers—entry level, junior scientist, mature professional, and "over fifty"—problems, adjustments, and joys.

To most of us already in the field, a career in science is better than almost anything else we might have chosen to do with our lives. We talk about making contributions to understanding, about solving problems that have previously defied solution, about association with bright productive colleagues, and about recognition of scientific achievements. We tend to talk less often about the demanding nature of the occupation and the many stress-inducing components of it, which are particularly characteristic of the early years of a developing career, but which never really disappear.

In this chapter, we have elected to dissect career pro-

225

gressions of women scientists into four marginally discernible phases:

1. Entry level

2. Junior scientist level

3. The mature professional

4. Women scientists over fifty

This designation of phases is obviously artificial, since a scientific career is usually a continuum, marked by peaks and valleys. It serves, however, as a convenient descriptive structure for the chapter, even though it collapses somewhat with phase 4—the older woman scientist—since in many instances categories 3 and 4 become indistinguishable chronologically. But these subdivisions may have some utility, if only as narrative hooks.

ENTRY LEVEL

Being an entry-level scientist is simultaneously an exhilarating and terrifying experience. Degrees have been earned but multiple, severe, selective forces are still at work, compounding those that may have already been experienced in graduate school. A major challenge is the first job—not just any job, but a *good* one, in a respected institution with compatible colleagues and opportunity for creative application of previous training. This is the dream, and for a few it becomes a reality. For most, though, the route upward is rougher, through endless and often fruitless submissions of résumés to major research universities, postdoctoral appointments as stopgaps, and servitude on the faculties of small colleges with inadequate facilities and salaries.

One of the most distressing aspects of this critical early period is the intrusion of chance events that can affect the recruiting process, and the realization that the entire future course of a career can be determined by some minor episode—positively or negatively—that will influence a hiring decision. Entry-level people must be virtual paragons if they are to be hired in today's sluggish academic and industrial marketplace. They must have superb training, strong support from a mentor, demonstrated research abilities, enthusiasm, and effective interpersonal skills. They must also match exactly the template of perceived needs of any employing institution. The wonder is that any hiring is done at all. Some recent job searchers report submitting in excess of 200 résumés.

Once the traumas of job interviews, pre-employment seminars, and inquisitions by the senior faculty have been survived and a position has been offered and accepted, the real selective factors determining upward mobility become operative immediately. Exceptional courses must be planned and given, overnight progress in research projects will be expected, grant applications must be prepared and then funded, faculty relationships will be expected to be cordial, students must be counseled with great diligence and insight, and seminars will be expected—all this in a maelstrom of personal stress associated with moving and getting established in a new location. Ah, those early years are great to look back on, but hell to live through!

If the institutional and personal demands are not enough, contact with and participation in the broader scientific community must be pursued vigorously. Papers are expected to be presented at society meetings and then published in peer-reviewed journals, seminars should be given at other institutions, committee assignments should be sought and then carried out enthusiastically, and networks of colleagues need to be established and then extended.

The word that best describes all this activity is overwhelming; but, good entry-level scientists do these things and, more importantly, do them well and with confidence. These are the incipient professionals who will be seen later in the choice roles in science.

It should not be unreasonable to ask at this stage in a daunting discussion on entry-level scientists the by-now ritual question: "How do gender considerations enter the processes described?" The best entry-level jobs—the plums—go to the best and the brightest from the best graduate schools, especially if there is aggressive support from a mentor who is "plugged into" the right networks of colleagues in a discipline area. This is reality and is only slightly mediated by factors related to gender. Outside the perimeter of the plum tree, other good positions exist; ones for which well-qualified junior scientists can compete successfully and in which they can prosper, with the right mix of ability and productivity. Here, being a woman scientist can be an asset or a liability: an asset if the institution has viable equal opportunity policies, or a liability if persistent, unstated, maybe even unconscious biases exist among those participating in the selection rituals.

Then, once selection is made, gender can continue as a positive or negative factor in advancement, depending on the same policies or biases. Our survey results indicate that the perception exists among at least some younger women scientists that scrutiny of their early performance is more intense than for males; that more is expected of them; and that no mistakes are allowed during this early period of employment. They feel that subtle and overt pressures to "prove themselves" are unusually severe, exceeding those felt by male cohorts.

One pervasive feeling, expressed variously by beginning female scientists, is *lack of collegiality and security*. They have just spent huge sums on graduate school tuition and are faced with the myriad problems of "living": cars that won't die on the road, apartments that are desirable, moving expenses that are not prohibitive. Will female scientists faced with such challenges be confident enough to insist on high salaries, perks, and rapid promotions, or will they tend to be passive, accepting whatever deal is offered?

JUNIOR SCIENTIST LEVEL: FAST FORWARD

The junior scientist phase of a career follows a course marked by successes and failures in executing a complex juggling act. It is a time when learning new professional and interpersonal skills must occur *simultaneously* with application of those skills to job performance. Scant allowance can be made for detours or derailments and few errors are tolerated. For the competent and energetic practitioner, it is a period when early rewards appear: funding of the first sizable grant application, publishing the first really substantive paper, being elected to office in a professional society, having the first graduate student complete a thesis, rubbing shoulders with international scientists, and meriting a raise. These rewards reaffirm that the choice of a scientific career was a correct one and that there really can be joy in science (along with some drudgery and frustrations!).

The junior scientist phase is also one of extraordinary

stress, especially related to balancing demands of professional and personal lives. This is particularly true for women scientists who elect to combine a scientific career with having a family. Maximum professional growth usually occurs coincident with peak reproductive years and major commitment to child rearing—an overwhelming assignment. The magnitude of these parallel demands can be for some women just too great; often a decrease in professional involvement and forward momentum is a consequence. Some institutional arrangements—liberal maternity and sick leave policies, provision of child care facilities, "stopping the clock" for tenure decisions—are designed to reduce the resulting competitive disadvantages, but they are only partially effective.

The response of one of our interviewees brings the dilemma into focus:

> In July I was offered the job of director of the marine laboratory for a Southwestern university. We agonized over our decision, but, as the move would have meant a cut in our combined income, we ended up saying no. Perhaps even more important than the salary question, however, was that with five weeks vacation per year, organized, government-supported daycare, days off (with pay) when the kids are sick, and six months maternity leave, it really is more feasible to combine career and family in Scandinavia than in the States.

The reality is that dual commitments—science and the home—can impose severe handicaps on the professional development of women scientists. But a number of women meet both challenges with considerable self-assurance.

What of the women scientists who don't choose to follow this dual path, who elect nearly sole commitment to their professional lives? This group of women scientists might be considered the "control" group in our study. Is their passage through this intense and highly selective junior scientist phase of careers modified by gender considerations? Are the joys, rewards, and frustrations comparable to those experienced by their male cohorts? Responses here are highly variable. Some point to continuing disparities in salaries, rates of promotion, and awards of tenure. Others find no disadvantages in being a woman, *provided* that their professional work and achievements are judged impartially (which they often are not). In fact, a number of respondents felt themselves to be *better* scientists than most of their male counterparts. We think that surveys (including ours) frequently emphasize very real and persistent inequities, but fail to give equal attention to the large proportion (in our case a majority) of women in science who state that they feel good about their careers and do not detect overriding competitive disadvantages at this phase of professional development.

Despite the many satisfactions of a career in science, some women find it highly stressful (as, of course, do some men). Scientists, almost by definition, are ambitious, self-critical, and uncompromising in demands on themselves as productive professionals. These characteristics, when combined with the increasingly competitive nature of scientific research, make stress an almost unavoidable consequence of a career in science. The litany of important contributors to mental pressures would be certain to include the necessity of ensuring a never-ceasing flow of research proposals and grants; the continuing requirement to keep up with the literature, to do meaningful research, and to publish significant findings; the never-ending need to prepare and present effective lecture/seminar material; and the constant worry about maintaining adequate career progress. The pace is frenetic.

Superimposed on these gender-free sources of stress are several that are particularly targeted at women scientists. Foremost among them would be childbearing and child-rearing responsibilities if the choice of the female scientist is to have a family. All these responsibilities at times can take their emotional toll. Also included in the list of "gender-augmented stressors" would be the spectrum of overt and subtle discriminations and putdowns that can be a persistent part of a woman scientist's existence. These annoyances can wear away at even the hardiest souls. Moreover, the twin fears of success or failure have been identified by sociologists as significant problems for female professionals. Some women scientists fear success as much as failure.

To give some idea of how stressful it can be to pursue a career in science, one need only look to a recent study of women chemists by Dr. Molly Gleiser.[1] She produced the shocking statistic that women chemists were five times more likely to commit suicide than were other American women, and that the proportion was far higher than for male chemists. Furthermore, the study indicated that 88 percent of the cases took their lives for job-related reasons (compounded by personal problems). Some of the reasons given were:

- The prospect of unemployment and the difficulty in finding grant support

- The intensely competitive nature of the profession— one in which the likelihood is high of encountering barriers to careers in the form of gender discrimination

• Difficulty in coping with professional success, especially if it is achieved at the expense of family stability and closeness

Now an examination of comparative suicide rates is admittedly too harsh a criterion to apply to career perspectives of women in science, but it is an interesting approach that may have applicability to men as well. A subsequent study of suicide among eminent *male* scientists[2] did disclose interesting differences: principally that non-work-related causes accounted for 88 percent of suicides, far greater than those due to work-related causes. This would indicate that men are not handling their public and private lives as well as had been expected. If work was all they cared about—as we have been given to believe—they wouldn't be committing suicide for non-work-related reasons. An alternative explanation would be that the sample population consisted of men who had achieved eminence in science, so work-related stress may not have been severe.

THE MATURE PROFESSIONAL

We listed some characteristics of successful professionals in Chapter 2, and a reexamination of that list indicates that it exemplifies mature scientists at the peak of their skills and productivity. At this stage scientists publish reviews and books, organize and chair symposia, head scientific societies, edit journals, present invited lecture series, and are recognized as authorities in their specialties. These are the true professionals, who have moved through the earlier phases of scientific careers with increasing distinction and recognition—models for all who enter the field. They are, of course, survivors of a severe competitive process, but, more than that, they are risktakers and achievers, who make significant contributions to knowledge in their chosen disciplines.

Scrutinizing this population of scientists in their advanced career phase, we see certain facts becoming apparent. Most obvious is that a large part of the membership is still male, a reflection of an earlier and still-persistent imbalance in the numbers of men and women in science. Beyond this very visible inequity, however, achievements have an individual rather than a gender basis.

Our survey embraced both male and female scientists, including a number of both genders who had made important contributions in their specialty areas. Internal criteria of worth that they applied to assessments of their careers were remarkably similar, with contributions to concept development and to understanding basic processes being among the most frequently mentioned. Comments by some women professionals were that in their estimation scientific competence and productivity should be considered as dominant criteria for success, to the total exclusion of gender. A few even stated that as they got older they ceased viewing gender as a significant factor in their success or failure in science—a view not shared by many other women in that or earlier phases of careers.

An interesting recent phenomenon seen in institutional treatment of mature women professionals has been the emergence of so-called "guilt promotions": the granting of long-overdue rewards of promotion and tenure to older female scientists who have made significant contributions in their fields. Motivation for such belated recognition is not always clear, but faculty pressures and activities of women's committees must be contributory. Our best example of this is seen in the recent history of Dr. Rose Torgay.

> Dr. Torgay was engrossed in her scholarly research for 35 years. While her work was well respected, her colleagues at her home institution continually devalued/undervalued her contribution, both to the discipline and to the institution. All committees appointed to review faculty for advancement/promotion bypassed recommendation for Dr. Torgay. Decade after decade she was turned down, and without a peep from her. When in the early 1980s the faculty and administrators were sensitized to an awareness of injustices in advancement of women, Dr. Torgay was advanced, not rapidly step-by-step, but from assistant professor to full professor with tenure.

Similar accounts were told to us about awards for women scientists. Here is one example:

> Dr. Colleen Kemp is an organic chemist associated with a small New England liberal arts college. Although the work environment was clearly chilly for her in the first few years, she persevered in the department, supporting her research on grants and contracts. During years that a department member went on sabbatical leave or when a member went to a scientific meeting or on vacation, Dr. Kemp was asked to substitute. She became popular with the students. Students were more comfortable seeking her assistance with class and laboratory material than they were with the professor in charge (even when she wasn't substitute teaching!). Dr. Kemp always had an open door and a congenial smile.
>
> Dr. Kemp took an active part in campus activities that might ease the path for future women in science. She was ridiculed, isolated, and teased by her department-mates for such actions. Her visionary research was well-recognized nationally and internationally, and her joys came from hands-on bench science and extrainstitutional recognition. Two years ago, a new department head was hired, one sympathetic to the special challenges that women face. The chairperson quietly nominated Dr. Kemp for

the alumni award for "outstanding faculty." The award carried a cash prize and a medal and major recognition. After a highly competitive process run by the alumni, Dr. Kemp was awarded the prize at graduation. The award was a total surprise to Dr. Kemp and her department-mates.

Dr. Kemp's comments to a friend following the award: "It has made all of the difference in the world. For the first time my colleagues seem to respect my self-determination and accomplishments. Finally, after nine years, I feel like I belong. I stand a little taller. I have been asked to assume a permanent teaching position and to lead the research for undergraduate programs. 'Acceptance' increased my productivity considerably. It is important that administrators and colleagues alike learn that 'icing' women rarely causes them to drop out. Instead, hyperdetermination or overdrive kicks in. When the work environment is sufficiently collegial and this overdrive can be relaxed because gender becomes a nonissue, productivity is increased. This alone makes working for absence of constraints worthwhile for males and females."

Our survey also dealt with women scientists in managerial positions, and our sample included some in senior decision-making as well as bureaucratic roles. Conversations with women in these more senior positions reaffirmed an observed extreme scarcity of females in key science administration positions (especially laboratory and institute directors and deans) and the possibility of lingering gender biases influencing the selection process. These comments reinforce the earlier perception that "women *do* science, but they don't usually *direct* it," through no fault of their own.

THE WOMAN SCIENTIST OVER FIFTY

It is important to point out early in the discussion of women scientists over fifty that we are talking about women who began their careers in the late 1950s and early 1960s, well before any major awakening of interest in improving the status of female professionals. Women's liberation and equal rights were concepts to flourish in the future. In the past, the female faculty populations of many institutions, in science as well as in other areas, were minute if they existed at all.

> Dr. Joan Ellison is an example of the female scientist who began her career during that period. She is a tenured Professor of Chemistry at a large Southwestern university. Now in her sixties, she has published continuously in one highly specialized technical area in which she is considered one of the half-dozen or so authorities. She prefers individual research; by choice her technician support has never exceeded one person throughout her entire career, and she is so demanding that she has had only a few—but a very bright few—graduate students.
>
> By the time of the upheaval in perceptions of the role of women in science—the 1970s in particular—Professor Ellison and most of the female scientists considered in this section of the chapter were well into their careers, and many of their cohorts had long since disappeared into the more traditional female occupations of home and child rearing. We are dealing, therefore, with highly selected *survivors* of a demanding, uncompromising, sometimes hostile system dominated by male scientists.

Perceptions about women scientists over fifty abound. A few of the more common ones, which we do not necessarily subscribe to, are listed here:

- Women scientists who are older are often totally dedicated to a niche—a narrow somewhat obscure area of research—in which they are frequently recognized authorities, with very little competition.

- Women scientists who are older often tend to be impatient with younger colleagues (especially the males), and often become openly critical of their backgrounds and competencies.

- Women scientists who are older tend to give few oral presentations at society meetings, preferring to communicate principally through their published papers.

- Many women scientists who are older are understandably disgusted and embittered by decades of discriminatory practices by institutional administrators—practices that have often interfered with their productivity and tarnished the pleasures of doing science.

Conversations with older women scientists, especially those who are approaching retirement age, reveal repeatedly a resignation or calm acceptance of existence in the male-dominated, often discriminatory system of science. Many have learned, over the decades, essential rules of survival in that system; surprisingly few express bitterness about it; and most seem reasonably or even fully satisfied with their chosen careers. (Of course, these women have been satisfied enough to remain in science. Those who weren't left long ago.) One respondent stated, "Women scientists in this age category either have resigned or are resigned to their fate."

These more senior women scientists do occasionally offer insights about survival, however. One nationally recognized botanist at a Midwestern university offered this advice during a symposium for female graduate students:

> "Yes, it is tough to get in, and once you are in you must work harder than the men do—produce more papers, teach better courses, and do research that really matters—but if you do these things you will find that ability and productivity are more

important than gender in determining progress in science." She then had to spoil the purity of this advice by adding the caveat that, "You may at times be given more credit than a man would get for doing something, just because you are a woman—and if so, accept it with a gracious smile."

In discussions with women scientists who are over fifty, the existence of chauvinistic deans or directors was frequently acknowledged. The almost universal attitude seemed to be that good productive female scientists can prosper despite such impediments provided they retain a sense of humor about implausible events and are willing to confront clearly discriminatory practices if they appear.

For some older women scientists withdrawal to an area of narrow specialization was a way to succeed and to emerge as an authority in that research area. This deliberate choice was a method of avoiding competition with large well-funded groups (usually headed by aggressive males) on the "cutting edge" of experimental science. It represented a route to achievement that did not require a large and often unavailable (to a woman) research infrastructure.

Analysis of the data for this section on women scientists who are over fifty has produced some findings that might not have been predicted by conjecture alone and that depart significantly from common perceptions. Notable is the absence of stated bitterness toward a system that has frequently blocked the full expression of their competencies, especially in the early decades of their careers.

Other findings are more debatable; they include these:

- Women scientists who are older tend to be consummate pragmatists, recognizing and accepting, at least outwardly, the continuing dominance of men in most scientific disciplines, but remaining isolated from the power structure of science.

• Women scientists who are older tend not to be risktakers; they more often follow an established research path in a narrow specialty in which they are considered authorities. (We have just seen, nonetheless, that there is a method to this approach.)

• Women scientists who are older tend not to be participants in team research, gravitating to individual studies in a specialty with which they are comfortable and which does not require major funding, since they've had trouble getting funding in their earlier days.

• Women scientists who are older tend not to be active participants in professional society affairs or in active networking with male or female colleagues.

• On a more positive note, older women scientists who have recently received their advanced degrees and entered the profession (often after raising their families) rarely have the battle scars or characteristics mentioned above.

Examining these conclusions about women scientists who are older, we see at least two problems with our analysis:

• Some older female scientists are clear refutations of every generalization that has been made in this section; they are aggressive, extroverted risktakers, fully involved in departmental concerns, maintaining close interactions with assistants and graduate students and occupying dominant positions in professional society activities.

- Most of the descriptions of older female scientists apply equally well to older *male* scientists and seem to be more intimately associated with the evolution of behavior with advancing age rather than with gender.

We are left then with a situation commonly encountered by those who would discern trends and propose broad generalizations—that variability in the available data is extreme, so retreat to the fascinations of individual case histories is a much more rewarding (and much safer) course of action.

SUMMARY

The career progression of a woman scientist usually represents a continuum which, for descriptive purposes, can be subdivided into four phases: entry level, junior scientist level, mature professional level, and "over fifty" level. Each of these intergrading phases has distinctive characteristics. The *entry level* is a time of intense preoccupation with finding the right job (or in some cases, any job). Once this major milestone is passed, a flurry of institutional and career responsibilities—teaching, committees, grant applications, research, publications—must be faced, often in the midst of personal upheavals. The second or *junior scientist level* is one of high mobility until the right job and work environment are found. This is a time when productivity and ability are critically important, but also a time when early rewards for excellence become apparent. The *mature scientist level* is one when scientists are at the peak of their skills and productivity, participating in the social infrastructure of science, and being recognized by colleagues. The *"over fifty" level* is not always readily distinguishable from the preceding level, except chronologically. It can be for many a period of continued high productivity and broadened perspectives.

Gender-influenced modifications or even disruptions of this continuum can of course occur at any point. Entry-level women scientists may encounter remnants of discriminatory practices and attitudes associated with hiring, and are certain to encounter some of the subtle forms of discrimination early in their careers. Junior scientists who are married come face to face with the competing demands of home and laboratory, sometimes to the detriment of one or the other, but sometimes not. Despite stressful aspects, a surprising number of women scientists are happy with their choice of career, and some even minimize the competitive disadvantages of being a female in a still predominately male enterprise. Satisfaction with career choice characterizes the mature professional woman, too, although some will admit to long-standing gender-based problems with male department heads and deans. These women are, of course, survivors of a severe selective process; many of them have not only survived, but have also prospered, often as leaders in their fields of specialization. Women scientists who are over fifty provide further examples of success and quiet satisfaction with an occupation in which the barriers to full female participation have been lowered substantially within their lifetimes, although not removed altogether.

REFERENCES

1. Molly Gleiser (cited by Malcolm Browne), in Women in chemistry; Higher suicide risk seen, *New York Times* (4 August 1987), p. 21.
2. Molly Gleiser and Richard H. Seiden, Famous researcher's ultimate stress: When doing science leads to suicide, *The Scientist* 4(23), 20 (1990).

CHAPTER 14

CONCLUSIONS AND A FORECAST FOR THE YEAR 2000

Better, or more of the same?

An almost overwhelming amount of data about the status of women scientists exists in hundreds of research reports and many books (most of them published by university presses). As is often the case, however, *interpretations* of the data and *conclusions* drawn from the data may vary markedly. This can be seen clearly in the matter of comparative salaries in which some authors emphasize the persistent gender-related differentials, whereas others (and we are included) point to the narrowing gap, especially at lower professional ranks. Similar variability in interpretation can be seen in the matter of causes of gender differentials in scientific productivity and publication. Some find a correlation with the demands of child rearing; others feel that the consequences of early acculturation can influence scholarly output; and still others find the multiple disadvantages of discrimination as obvious causes. Reality undoubtedly resides in some central zone within the range of current statistical analyses, but not fully detectable by even detailed probes and therefore subject to wide variations in interpretation.

Searching diligently for a simple descriptor for the findings discussed in this book, we have identified "cautious optimism" as one leading contender. As with many social phenomena, gradual positive change, interrupted by plateaus of lessened activity, is the most likely course of the movement of female scientists toward equality. Much of the decade of the 1980s can be discerned as such a plateau, even though some improvements during this period can be detected. The record for the years 1980 to 1990 is erratic. The salary gap between male and female professionals seems to be closing, but a gap still exists; a leveling off has occurred in the numbers of women entering science and in the proportion of women on university faculties; and only slight improvement can be seen in the numbers of women at higher professional ranks and in involvement of women in international science endeavors. Women are appearing with somewhat greater frequency in positions of power and influence in science, but their numbers are still disproportionately small.

Changes in attitudes are more difficult to discern. Few near-term (the past ten years) changes in attitudes of male scientists about female scientists can be detected. With many males, there is still fear of the very competent woman scientist, combined with a deep-rooted feeling of superiority and inability to accept females as peers. Many, if not most, junior level female scientists classify themselves as "feminists"—either active or passive—and the feeling persists among them that women are not encouraged to join the enterprise and in some instances are actively encouraged to leave. Others are more sanguine about their status and prospects for the future, but are convinced that the system of science must be pushed and that women must be retrained away from passivity—that elimination of patterns of discrimination "will not come from a surge of male benevolence," as aptly put by Anne M. Briscoe.[1]

We think we have seen, from the results of our interviews, a gradual "lightening" in the working relationships of men and women scientists during the 1980s, as compared with the intense partisanship of the 1970s when gender-based battle lines were more clearly defined. The "comfort level" between male and female professionals seemed to increase in the 1980s, even though skirmishes still occurred; we assume that this trend will continue in the 1990s and beyond, despite some islands of resistance among a few male colleagues.

Our study has reaffirmed that women scientists' highest priorities are to advance ideas and knowledge, rather than to concentrate on rewards of dollars or recognition. However, this generalization should in no way obscure the need for equal access to all career benefits—tangible as well as intangible—regardless of gender.

Scientists come equipped with their human share of neuroses—even incipient psychoses—in a profession with informal as well as formal hierarchies, unwritten and sometimes poorly understood game rules, personality conflicts, and power struggles. *Survival* is important, as well as success. Any factors that affect the ability to do good science can be of significance; here we have examined aspects of the gender factor, of being a woman scientist in a male-dominated community of professionals.

A superb earlier study of women in science by Jonathan

R. Cole[2] explored in great depth the concept of three prominent historical barriers confronting the women who would
be scientists, barriers that Cole labeled the "triple penalty."
They are:

1. A cultural definition of science as an inappropriate
   career for women

2. A belief that women are less competent and less
   creative than men, and

3. An actuality of discrimination against women in
   science.

These barriers, according to Cole, helped to create
"marginal women" in science: women trying to live in two
worlds, seeking full participation in a male-dominated and
resistant scientific community, and still being evaluated
partly on the basis of fulfilling traditional female norms of
marriage and family.

During the past two decades the barriers have become
less formidable or, in some instances, have disappeared
altogether, and marginality for women scientists has been
replaced by a condition often approximating equality. Cole
summarized some of the elements involved, including dis-
avowal of the idea that women are incapable of doing creative research, acceptance of women into the informal structure of the scientific community, absence of exclusion of
women from scientific forums and organizations, and emergence of women in positions of power and prestige in
science. He pointed out, however, that all aspects of the
"triple penalty" have not vanished; that the quality of the
graduate experience for women may still be deficient, and
that barriers to promotion to high rank still persist. A final

assessment by Cole was that gender discrimination in science had diminished significantly by the late 1970s. That would also be our general conclusion at the end of the 1980s, based on findings from our study.

Our conclusion is somewhat more optimistic than that reached by Natalie Angier[3] in a recent *New York Times* article based on an extensive series of interviews with leading women scientists and incorporating data from the National Research Council. Among the most significant of her findings are:

- Women have become more numerous in the lower and middle echelons of science, but they are still inadequately represented in the upper strata—full professors, department heads, program leaders, laboratory directors, members of the National Academy of Science—and salaries remain disparate (Figure 14 and Table 2).

- Science is a gregarious occupation, often requiring team efforts, and men feel more comfortable collaborating with other men. This, for women, reduces the social interactions during which important information is transferred and places women at a disadvantage.

- Science has become more and more international, and male professionals from "conservative" cultures may be even more reluctant to collaborate with women than are American males.

- Women remain relatively invisible when choices are made by male conference and symposium organizers—possibly through oversight or "benign neglect."

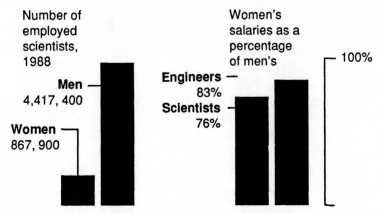

FIGURE 14. Relative numbers of employed female and male scientists in 1988; women's salaries as a percentage of men's in 1986. SOURCE: Natalie Angier, 1991.[3]

TABLE 2
TENURE/NONTENURE TRACK AND ACADEMIC RANK FOR PH.D.s, 1987[a]

	Women	Men
Tenure track	58%	74%
Tenured	36	60
Nontenured	22	14
Nontenure track	18	7
Other and no report	24	19
Academic rank		
Full professor	18	7
Associate professor	25	24
Assistant professor	29	15
Other and no report	28	15

[a]From Natalie Angier, 1991.

And when women are included on the program, they are subjected to closer scrutiny and greater criticism.

- Many women are unwilling to join the often brutal infighting that can be a part of scientific "discussion," whereas many men seem to expect and even enjoy rude bare-knuckled give-and-take.

It seems obvious that finding a common ground among various interpretations of the available information—a balance between our "cautious optimism" and the "more of the same" observations of Ms. Angier[3]—is difficult and maybe a little frustrating. In this book we have tried to augment the data base and have offered our conclusions and opinions. We think that the status of women scientists has improved significantly, but we do not discount the inequities that still exist.

REFERENCES

1. Anne M. Briscoe, Roadblocks remain for women scientists, *Chemical and Engineering News* (5 March 1984), pp. 30–31.
2. Jonathan R. Cole, *Fair Science: Women in the Scientific Community* (New York: Free Press, 1979), pp. 255–257.
3. Natalie Angier, Women swell ranks of science, but remain invisible at the top, *New York Times* (21 May 1991), pp. C1, C12.

EPILOGUE

Is it time to change the rules by which the games of science are played? Many women scientists believe that it is. There can be a better ethos. When we state this, what do we mean? Ethical behavior connotes honesty, justice, compassion, caring, nurturing of others. Certainly it can be effective as the traditional power play of "king of the mountain" and deserves to be tested.

A question was posed at a recent forum of women oceanographers at a national oceanographic meeting (The Oceanographic Society, February 1991): "Would the discipline of oceanography be different today if women had even co-leadership during the past 100 years?" The response by women was divided. Some quickly responded, "No, searching for the truth is gender free." Others, responding with less assurance and after considerable thought, "Perhaps." The lively discussion evolved to include thoughts centered around two words, *cooperation* and *competition*. Cooperation versus competition has changed the way we do science, as well as the subject matter of our science.

Yes, the way in which science was approached might have been different. Instead of emphasis on competition among scientists for research grants (there is even a program

within the National Science Foundation "to increase competitiveness"), shiptime, publications, travel awards, and invited symposia speakers and chairpersons, the emphasis
would have been on teamwork and cooperation. Is it now
time for a special program "to increase cooperation"?

Yes, the subject of scientific inquiry might have been
different. The paradigm on which biological oceanography
has been based for the past 100 years is competition based.
Who eats who? Who outcompetes whom for nutrition,
habitat? The paradigm is the vicious conflict of eat or be
eaten—survival of the most competitive. Little by little this
paradigm is receiving a fresh new look, a fact for which
women oceanographers would like to take partial credit.
Women oceanographers, by persistent attention to detailed
research that features the worth of the individual cell, the
individual organism, cooperation of various cells and organisms, have convinced others that the subtle symbiotic
associations—population synergy to advance the well-
being of the species, indeed, cooperation—can be as much
a selective advantage as is the ability to outcompete.

CMY

Thumbing idly through the pages of this book I noted
with some chagrin a distinct paucity of humor in most of
the chapters. Early consternation gave way quickly, however, to the sensible rationalization that the subject of
women in science is a serious one and should be presented
as such. It seems inappropriate to treat with levity a subject
that is so important to science and especially to women
scientists, for whom the historical reality of unequal professional treatment contains little humor. Our optimistic conclusions about a perceptible improvement in the status of

women in science should be an occasion for some joy, though, even in a less than perfect world.

One no doubt fascinating aspect of the kind of authorship that this book has had—a woman and a man trying to write jointly about women in science—is the psychological distance that separates the two writers and the inherent question about whether the arrangement can possibly lead to any kind of consensus. From the male perspective (CJS) the path has been remarkably smooth, once a few of my nasty preconceptions about women scientists were exposed and eliminated forever. Some of those preconceptions had crept into my earlier books about scientists, so proper atonement brought a measure of relief, as well as an opportunity to examine aspects of the real world of the woman scientist. I would recommend a similar exercise for any male scientist, except that the publishing world can absorb only a limited number of books on the topic, regardless of its fascination.

Clarice and I have labored intermittently for four years, through peaks and valleys of enthusiasm and despair, to complete this book; if our editor, Linda Greenspan Regan, had allowed us four more years, it might have been that much better and more exhaustive, but we think it is in reasonable condition right now, without further tampering or manipulation.

CJS

SPECIAL PROGRAMS AND ORGANIZATIONS SUPPORTING WOMEN CAREER SCIENTISTS, MATHEMATICIANS, AND ENGINEERS
A Sampling

American Chemical Society of Women Chemists
1155 Sixteenth St., NW
Washington, D.C. 20036
(202) 872-4600

Association of American Colleges
1818 R St., NW
Washington, D.C. 20009
(202) 387-3760

Association for Women in Science (AWIS)
1522 K St.
Suite 820
Washington, D.C. 20005
(202) 408-0742
FAX (202) 408-8321

Association of Black Women in Higher Education
30 Limerick Drive
Albany, NY 12204
(518) 465-2146

Committee on the Career Advancement of Minorities and
   Women
Council for the Advancement and Support of Education
   (CASE)
11 Dupont Circle, NW
Suite 400
Washington, D.C. 20036
(202) 328-5930

Committee on Women in Science and Engineering
National Academy of Sciences
Harris Building
2001 Wisconsin Ave., NW
Washington, D.C. 20007
(202) 334-2000

Department of Education
Women's Educational Equity Act Program (WEEA)
Division of Discretionary Grants
400 Maryland Ave.
Washington, D.C. 20202
(202) 401-0351

National Association for Women Deans, Administrators and
    Counselors
1325 18th St., NW
Suite 210
Washington, D.C. 20036
(202) 659-9330

National Science Foundation
1800 G St., NW
Washington, D.C. 20550
    Faculty Awards for Women (202) 357-9639 or
        (202) 357-7461
    Research Initiation Considerations, Research Planning
        Grants, and Career Advancement Awards (202)
        357-7456
    Visiting Professorships for Women (202) 357-7734

American Association for the Advancement of Science
1333 H St., NW
10th Floor
Washington, D.C. 20005
(202) 326-6680

Science-by-Mail™
Museum of Science
Science Park
Boston, MA 02214-1099
(800) 729-3300 or
(617) 889-6211

Women's Aquatic Network, Inc.
P.O. Box 4993
Washington, D.C. 20008
(202) 226-2460

Women's Committee Council of Graduate Schools in the U.S.
Dean of Graduate Studies and Research
Appalachian State University
Boone, NC 28608
(704) 262-2130

THE WOMAN SCIENTIST QUESTIONNAIRE

Highest degree:	☐	BS
	☐	MS
	☐	Ph.D.
Gender:	☐	Female
	☐	Male
Age group:	☐	< 25
	☐	25–29
	☐	30–34
	☐	35–39
	☐	40–49
	☐	50–65
	☐	> 65

Scientific discipline: _____

Section I. To be answered by male and female scientists
1. How would you define "success" in science?
2. What are the things about "doing science" that you find to be most satisfying?

3. How would you describe your scientific accomplishments so far in your career?
4. Are you satisfied with your rate of career progression?
5. If you have daughters, have you or will you encourage them to have a career in science? Why or why not?
6. How do you see the future of women in science—ex. year 2000?
7. Have you had a mentor relationship with a male or female senior colleague? If so, can you describe it?
8. How would you describe your record in obtaining grant support from outside agencies? Which agencies?
9. What has been your experience with female science managers, and what conclusions can you draw from it? Which agencies?
10. How do you relate to the "in-groups" or "scientific clubs" in your discipline?
11. Have you seen changes in the past decade in the roles of women scientists? If so, what are they?
12. Do you believe that equality is being reached in the treatment of female scientists by administrators?
13. In team or cooperative studies, do you prefer to work with male or female leaders? Colleagues? Technicians? Why?
14. Do you supervise or manage a research group, and, if so, what are the strengths and weaknesses in your performance?
15. Are women scientists accepted as "full partners" in peer groups in your discipline and your institution?
16. If you have graduate students, are your graduate students mostly male or female? Why?

17. Do you feel that you have made a significant contribution to concept development in science? If so, what?

Section II. To be answered by female scientists

1. Would you describe yourself as a "feminist"? (active, militant, passive, anti?)
2. Do you feel that the feminist movement of the 1970s has had an impact on your scientific career? If so, how?
3. How would you describe your relationships with other female scientists of similar career status?
4. How would you describe your relationships with male scientists of similar career status?
5. Do you *feel* equal to male cohorts—insofar as treatment by director (department chairperson), promotions, travel, etc., are concerned?
6. Have you been part of or know of examples of female sabotage to females?
7. Have you ever felt like the "invisible woman" at seminars, conferences, small-group discussions, staff meetings, etc.?
8. Have you ever experienced the "impostor syndrome" in your own career?
9. Are you most important contacts male, female, or both?
10. How closely does your salary approximate that of male counterparts in your institution at the same career stage and with similar abilities?
11. Have you encountered any or all of the following discriminations:
    • an "excluding hierarchy" of male colleagues whose existence and control are denied?

- the perceived need to prove oneself constantly—far beyond that expected of a male?
- the difficulty in obtaining promotion, tenure, and grants—compared to men?

12. If you have felt discrimination, what has been the effect on your scientific creativity and productivity?

13. Is the "research associate syndrome" for female scientists still alive and prospering at your institution? (The syndrome is best characterized by the availability of nontenured research associate positions and the scarcity of tenured faculty positions.)

14. If not married, or coupled, have you ever experienced any pressure or desire to be so? If so, what was the source of the pressure?

15. How would you describe your partner's attitude toward your career? (highly supportive, supportive, unsupportive)

16. If you do not have children, have you ever experienced any pressure or desire to have a child/children?

17. If you have children, have you encountered:
   - appropriate maternity leave?
   - "stopping-the-clock" for advancement/tenure decision?
   - pressure to return to the job before you were ready?
   - appropriate workloads/flexible schedules upon return to permit nursing?
   - adequate child care facilities
   - sick leave granted when children ill?

18. How would you describe geographic flexibility with your career pursuits? (mobile; mobile, but tied to partner; immobile; willing to live separate from partner)

Index

Printed in the United States
50331LVS00006B/11

9 780738 208824